SURVEYS OF FISHERIES RESOURCES

SURVEYS OF FISHERIES RESOURCES

Donald R. Gunderson

University of Washington, School of Fisheries

JOHN WILEY & SONS, INC.

New York / Chichester / Brisbane / Toronto / Singapore

This text is printed on acid-free paper.

Library of Congress Cataloging in Publication Data:

Gunderson, Donald R. (Donald Raymond), 1942–
Surveys of fisheries resources / Donald R. Gunderson.
p. cm.
Includes bibliographical references (p.).
ISBN 0-471-54735-2 (cloth)
1. Fishery resources. 2. Fish stock assessment. 3. Fishery management. I. Title.
SH327.5.G86 1993
333.95′611—dc20 92-21522

Printed in the United States of America

10 9 8 7 6 5 4 3 2

PREFACE

Direct surveys, where ichthyoplankton nets, trawls, acoustic systems, or visual counts are used to determine the abundance and age composition of a given fisheries resource, have been a common research and management tool for decades. The use of such surveys has become increasingly widespread in recent years, however, as fisheries resources have become exploited more intensively (overexploited in many cases), and require frequent, accurate monitoring if they are to be managed properly. The methodology used in these surveys has improved rapidly to meet the demand for accurate, up-to-date information on the status of fishery resources, and has become increasingly complex.

Despite the fact that all survey methods share common elements in terms of statistical design, their inherent technological differences have led to a certain degree of specialization. Information on resource surveys is consequently scattered throughout a broad range of publications, making it difficult for a beginner to become familiar with the basic requirements for an effective survey, and to evaluate alternatives in technology, methodology and survey design.

The purpose of this book is to integrate the most important aspects of fisheries surveys in a single volume. A balanced approach is taken to the general problem of estimating the abundance and age composition of fisheries resources, with equal emphasis on consideration of the biology of the target species, selection of appropriate survey technology, and statistical design of the survey. The assumptions and limitations inherent in each survey method are discussed, and examples of successful applications are provided. Care has been taken to cite pertinent references throughout the book, so that the reader can refer to them when a greater level of detail is desired.

This book is intended for graduate students and upper division undergraduates with an understanding of basic statistics. It should also prove useful to biologists, ecologists, and statisticians involved with the problem of monitoring the abundance of fish, invertebrates, or marine mammals in the field.

DONALD R. GUNDERSON

Seattle, Washington
January 1993

ACKNOWLEDGMENTS

I wish to thank all authors and editors who gave permission to redraw or reproduce published material, and all colleagues who provided me with unpublished illustrations. All original drawings were done by Yun-Bing Shi, whose assistance was indispensable throughout the preparation of this book. Several colleagues reviewed early drafts of certain sections, and I would like to thank all of them for their help. I am particularly grateful to those who reviewed entire chapters: John Hunter, Art Kendall, Dan Kimura, Nancy Lo, Beverly Macewicz, Ed Nunnallee, Jim Traynor, Bill Karp, Bill West, Al Erickson, and Waldo Wakefield. I also wish to thank Mauré Gunderson, Marcus Duke, Abby Simpson, Cathy Schwartz, and all other staff members at the University of Washington for their technical assistance.

CONTENTS

INTRODUCTION

Estimates of abundance for fish populations have always been required for fisheries management purposes, and the demand for such information has increased dramatically in recent years. Population estimates or indices are required before any real quantitative understanding of ecological processes is possible, since estimates of abundance, mortality rates, and recruitment are an integral part of most studies aimed at determining the processes that regulate population size. Abundance surveys will always play a major role in evaluating the relative importance of environmental influences, interspecies interactions, and intraspecific density-dependent effects in regulating population size. In many ways, the accuracy with which population sizes can be determined ultimately defines our ability to test ecological hypotheses in the field.

A number of monitoring programs have been instituted to evaluate the impact of pollution, habitat degradation, offshore oil field development, construction of power plants, etc. on fish populations. The ever-increasing number of "recovery plans" for endangered species offer further examples of programs that require careful monitoring of population sizes. These programs all require careful experimental design, and abundance surveys with considerable statistical precision if they are to function properly. In fact, monitoring programs form only a subset of broader ecological studies, since there is a need to separate the effects of the management activity in question from natural events that occur in the ecosystem. The problem is a complex one, and there can be little doubt that requests for accurate surveys of abundance will continue to accelerate in the future.

Until the 1960s, most stock assessment relied on analysis of catch per unit effort (CPUE) data from commercial fisheries to analyze trends in the abundance of fish stocks. In recent years, use of CPUE data has declined considerably owing to a number of problems, the most notable of which are the following:

1. CPUE is difficult to define in multispecies fisheries or where search time is an important part of fishing operations (Gulland 1983).

2. Innovations in commercial fishing technology are often rapid, with several occurring simultaneously. As a result, and because individual fishermen differ radically in their effectiveness, statistical adjustments to correct for changes in the "standard" gear are usually difficult to make. As an example, Pacific halibut (*Hippoglossus stenolepis*) CPUE data employed by Quinn et al. (1985) required adjustments for hook spacing, hook type, seasonal changes in catchability, and regional effects before they could be used in an analysis of stock production relationships.

3. The basic assumption that there is a simple, linear relationship between catch per unit effort and stock size (CPUE $= qB$, where $q =$ the catchability coefficient and $B =$ stock biomass) has been shown to have little basis in fact. Recent studies have shown that q is actually inversely related to population abundance, following a relationship of the form $q = aB^{-b}$. (Fig. 1) This results from the fact that fishermen are able to locate areas of high local abundance even when overall stock sizes are low, and concentrate their fishing effort there. This has been shown to be the case for a wide variety of fisheries, including purse seine fisheries for sardines (McCall 1976), anchovies (Csirke 1988), and herring (Ulltang 1980), trawl fisheries for cod (Garrod 1977) and haddock (Crecco and Overholtz 1990), as well as gill net and longline fisheries for cod (Angelson and Olsen 1987).

4. Widespread use of catch-at-age analyses such as virtual population analysis (VPA) and cohort analysis began in the 1970s, allowing estimates of population size and stock biomass to be made from data on the age composition of the catch. Once these estimates became available, it was no longer necessary to rely on trends in CPUE as proxies for trends in stock biomass.

5. Improved navigational and positioning techniques, together with the development of quantitative sampling methods, have resulted in widespread use of direct surveys to estimate stock biomass. Estimates of stock biomass from the data obtained during a single trawl, acoustic or ichthyoplankton survey are now common, without the time lags inherent in analysis of CPUE or catch-at-age data. Results from these surveys are not influenced by misreporting of commercial catch data or by errors that result from erroneous age determination.

Improvements in both catch-at-age analysis and direct surveys have continued, and they are now indispensable components of any modern stock assessment program. The assumptions underlying both catch-at-age and survey estimates of biomass can lead to significant bias, however. Catch-at-age

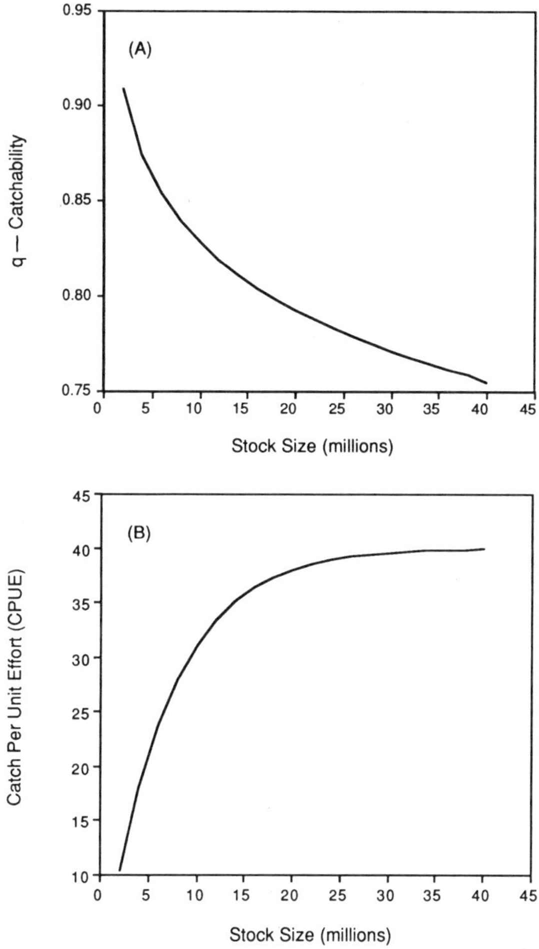

Figure 1. Density-dependent change in *q* commonly seen in commercial fisheries (A), and the resulting nonlinearity between stock size and catch per unit effort (CPUE) that occurs as a result (B).

analysis is subject to errors in the basic accuracy of the catch-age matrix owing to misreporting of catches or age determination errors, and bias can also enter through the choice of natural mortality (M) or terminal fishing mortality (F) coefficients used to initialize the analysis (Ulltang 1977; Bradford and Peterman 1989). Several researchers have shown that catch-at-age data alone are insufficient to estimate absolute abundance, and a variety of methods have been developed to incorporate auxiliary data on fishing effort (Pope and Shepherd 1985) or direct surveys into the analysis in order to stabilize the results. Nevertheless, significant bias can still result when erroneous biomass estimates from direct surveys are incorporated into the catch-age analysis

(Kimura 1989), and the results obtained are heavily dependent on the accuracy of the surveys (Rivard 1989).

Direct surveys of abundance have their own sources of intrinsic bias, and as a result independent verification of any biomass estimate using alternative techniques should be an integral part of any stock assessment program. Budget constraints often make this impossible, but biomass estimates for a valuable resource are often based on at least two independent estimation techniques. This will usually be sufficient to expose the most important sources of bias inherent in each technique, although a third estimate may be desirable in some cases.

This book will focus on population (or biomass) estimates employing direct surveys that require only a single pass through the study area. Other approaches to population estimation, such as catch-at-age analysis or tagging experiments have been reviewed elsewhere (Ricker 1975, Megrey 1989, Hilborn and Watters 1992), and the reader is referred to these sources if further information is desired.

The book will begin with a discussion of biological and statistical considerations involved in designing an effective survey. The sampling equipment and methodology used in conducting direct surveys will then be examined, with equal attention being given to trawl, acoustic, ichthyoplankton and sighting surveys. Alternative approaches used in each type of survey will be reviewed, and case histories used to demonstrate successful applications.

1

SURVEY DESIGN

BIOLOGICAL ASPECTS

The most important step in developing a direct survey of abundance is to obtain a complete understanding of the distribution of the target species, and the effective sampling area of the gear being used to census it. All of the survey methods described in this book are based on the simple extrapolation of the number (or weight) of animals observed per unit area, using a given sampling gear, to the entire survey area. All estimates of abundance can consequently be placed in the conceptual framework of quadrat sampling, i.e., $\hat{B} = A\overline{D}$, where $\hat{B}$ = biomass estimate (g), A = total survey area (m^2), and $\overline{D}$ = mean density, or weight observed per unit area (g/m^2) using the sampling gear. Alternatively, an estimate of population size ($\hat{P}$, numbers) can be obtained simply by defining $\overline{D}$ in terms of number observed per unit area. In this context, it becomes clear we are implicitly assuming that:

1. The entire population of the target species is in the survey area at the time the survey is being conducted, and will be censused only a single time (availability coefficient = 1.0).
2. The target species is completely vulnerable to the sampling gear being employed,with no gear avoidance or size selectivity.
3. The area sampled by the gear is known exactly.
4. The sites selected for sampling are a random, representative selection, and the mean abundance at these sites can be extrapolated to the entire survey area.

Availability

The first of our assumptions requires that the survey area completely brackets the geographic distribution of the target population. The first year of research for any survey should allow ample time for exploration, so that the geographic and bathymetric limits of the target species can be carefully defined. It is usually necessary to sample well beyond the regions where the fish are present, to be certain that the distribution of the target species has been adequately mapped out. If this is not done prior to the establishment of the routine survey grid, the investigator runs the risk of either failing to sample the entire range of the target species, or wasting valuable time surveying in areas where they do not exist. Since interannual variation in distribution patterns may occur, it is usually a good idea to survey beyond the bounds of the routine survey from time to time in order to be certain that the entire stock is being covered. This approach has sometimes been carried even further in the case of acoustic surveys, with a fixed fraction of the total survey effort being allocated to a preliminary "outline survey" that is used to delineate survey boundaries and allocate sampling effort.

Our second assumption requires that all individuals in the survey area are vulnerable to the sampling gear being used. Since any sampling gear has characteristic "blind spots" where it is incapable of detecting fish, we must assume that the behavioral characteristics of our target species are such that no individuals are present in these areas. In the case of trawl surveys, for example, all individuals are assumed to be present in the near-bottom area defined by the height of the trawl headrope. This is warranted for demersal species such as flatfish or crab, but leads to considerable bias in the case of semidemersal species such as Pacific hake (*Merluccius productus*) or walleye pollock (*Theragra chalcogramma*).

It is often possible to meet the apparently stringent requirements of this assumption through a careful choice of the season and time of day that the survey is conducted, locating a "window of opportunity" where the target species is completely available to the sampling gear being employed. Availability can vary substantially with time of day, with most species of fish showing distinct diurnal migrations. This typically involves a migration to the upper portion of the water column at dusk, and a return to the near-bottom zone at dawn. Migrations of this nature are partially responsible for the fact that bottom trawl catch rates for redfish (*Sebastes* spp., Atkinson 1989), haddock (*Melanogrammus aeglefinus*, Engås and Soldal 1992), and cod (*Gadus morhua*, Pitt et al. 1981, Fig. 1-1) are higher (3-fold and greater) during the day than at night. Migration patterns often change with size and age, resulting in significant diel differences in the size composition of the catches as well. Day–night intercalibration coefficients can be calculated to adjust for these effects statistically (e.g., see Chapter 2, Gear Standardization and Intercalibration), but the high variances associated with them often make them unreliable. Under these circumstances, it is best to restrict trawl sampling to the daylight hours.

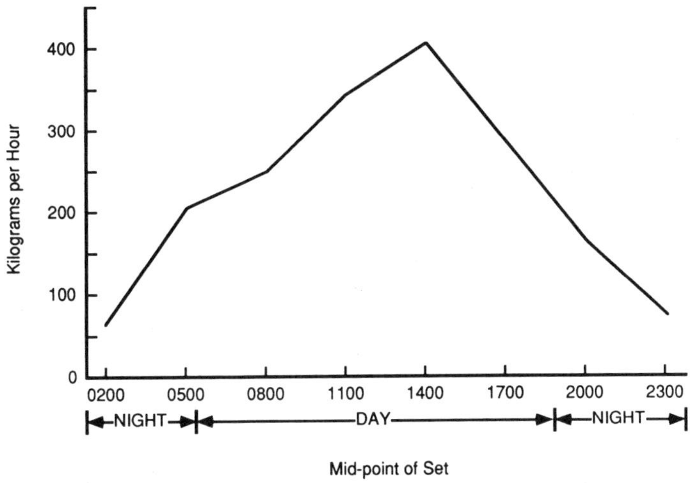

Figure 1-1. Diel variability in the average catch per hour for cod on the Grand Banks (Pitt et al. 1981).

A contrasting situation holds for many acoustic surveys, since individuals located near the bottom may not be detectable, making it necessary to conduct the surveys at night. As a result, nighttime acoustic surveys are the norm for Pacific herring (*Clupea harengus pallasi*), sockeye salmon (*Oncorhynchus nerka*) smolts, and a number of other species.

Seasonal migrations are also a common feature of the life history of most species, and it is only rarely (bivalves, highly territorial fish, etc.) that the target species can be assumed to be stationary. Unless the investigator has a clear idea of the migration routes and migratory behavior of the target species, and designs the survey in such a way that migration effects are accounted for, considerable bias can result. If the survey progresses slowly in relation to the migration of the target species, it is possible for some individuals to be counted several times (survey direction and fish migration proceeding in the same direction) or elude detection (survey direction opposite that of the fish migration). In cases of this nature the effects of migration can be offset to some degree by adding a sufficient number of survey vessels to obtain a synoptic "snapshot" of the target population over a restricted time period.

Some of the least usable survey data result when the population being surveyed bears an uncertain relation to the total population. Counts and size estimates for schools of fish (herring, sardines, etc.), where there is no clear idea of the proportion of the total population that is actually schooling, offer one clear example. Only by assuming that the proportion of fish located in detectable schools is constant over time (clearly unjustified in most cases) and treating the results as an "abundance index" can the data be of any use.

Surveys of spawning fish, where the survey takes place in the spawning area and censuses some unknown portion of the total population as individuals enter and leave the area being surveyed, offer another example of survey results that lend little insight to the actual abundance of the target population.

Sampling Gear

A third requirement for quantitative surveys of abundance is that the area being sampled by the survey gear is known exactly. This is more difficult than one would imagine, and few surveys meet this requirement completely. The employment of quadrat frames of known size to sample bivalve populations offers an exception to the rule, but it is usually much more difficult to determine the actual sampling area (or volume), even for commonly used "active" survey gear such as trawls, plankton samplers, or acoustic systems. Beam trawls may fail to tend bottom adequately, and plankton samplers can become clogged, even though the physical dimensions of the mouth opening have been fixed in both cases. The volume being sampled by an acoustic system at any given time is dependent on system parameters, background noise, and the average target strength of the species being sampled.

In the case of passive sampling gears such as gillnets, longlines, pots, and traps, the inability to define the sampling area precludes direct estimation of abundance. While efforts have been made to develop a methodology for estimating the effective area sampled by such gear types (Eggers et al. 1982), they have relied on somewhat simplistic assumptions. The most notable of these is that the capture field around each unit of gear can be treated as a circle with a constant radius. In fact, the capture field is known to vary with a number of factors such as prevailing currents, activity levels of the target species, and attractiveness of the bait. A further problem lies in the manner in which passive gears tend to become "saturated" at some point (Fig. 1-2), so that the effective sampling area actually varies over time. Surveys with passive gears can only be expected to provide a relative index of abundance, and even then it must be assumed that the capture field remains fairly constant between surveys.

Purse seines and midwater trawls are almost always deployed on schools of fish that have been located through extensive searches with acoustic equipment or aircraft. Random fishing sets are seldom made with these gears since the schooling pelagic species on which they are deployed are usually clustered in a very small portion of the survey area. While it is sometimes possible to define the surface area occupied by schools during preliminary aircraft or acoustic surveys (Burnham et al. 1980; Wilkins 1986; Hara 1989), quantification of school biomass and the proportion of the target population that is concentrated in schools at any given time is usually problematic (Hewitt et al. 1976). As a result, the use of midwater trawls and purse seines in fisheries surveys is usually confined to the identification of species and size composition of aggregations that have been censused during systematic acoustic surveys.

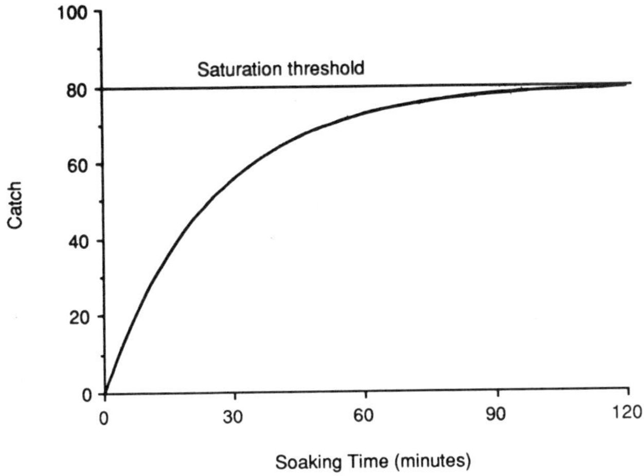

Figure 1-2. Relation between catch and time fished (soaking time) for passive gears.

Visual counts of individual organisms (fish, marine mammals, crabs, etc.) or schools are commonly made by divers, remotely operated vehicles (ROVs), and aerial and shipboard observers. The area being effectively sampled in these instances can usually be quantified by rigidly defining a strip of known width and counting only those individuals within this area. Alternatively, the observer can follow a set transect line, carefully recording the distance off-transect that each animal is detected. The effective area sampled can then be determined empirically using a number of statistical techniques (Burnham et al. 1980). Implicit in all direct counting surveys, however, is the assumption that the observer is not influencing the distribution or behavior of the target species.

Vulnerability and Selectivity

Vessel Avoidance

Many species react to light or noise generated by survey vessels, submersibles, etc., and can potentially initiate avoidance responses well in advance of the observer or sampling gear. For example, Buerkle (1977) estimated that cod are capable of detecting the noise generated by a small trawler at a distance of 2.5–3.2 km. As such, it is impossible to evaluate the full range and impact of escape behavior without using long-range sonar to scan the area ahead of the survey vessel, or a second observation platform that is not detectable by the target species.

Misund (1990) used a multibeam, sector scanning sonar (Chapter 5) to make observations of herring schools, and observed diving by the schools only when the vessel passed directly over them. However, the schools avoided the vessel horizontally in about 70% of the cases he studied. Radial swimming directions relative to the survey vessel can be readily determined using a multibeam sonar, and behavioral observations for a number of schools of herring (*Clupea harengus harengus*) and sprat (*Sprattus sprattus*) (Misund and Aglen 1992) have shown that many schools are herded in front of the vessel while others avoid the vessel by swimming to either side of it (Fig. 1-3).

This herding behavior appears to be a response to the noise directivity pattern produced by the vessel, and implies that the fish sense an increase in sound intensity when coming into the main lobes of the vessel noise, and react by swimming toward the forward path of the vessel where the noise intensity is lower (i.e., toward the "tunnel" of low intensity sound in front of the vessel). The typical pattern observed in front of the vessel was forward–sideways–forward swimming, as if the school was keeping itself in the sector of minimum sound intensity. This herding effect appeared to be most prev-

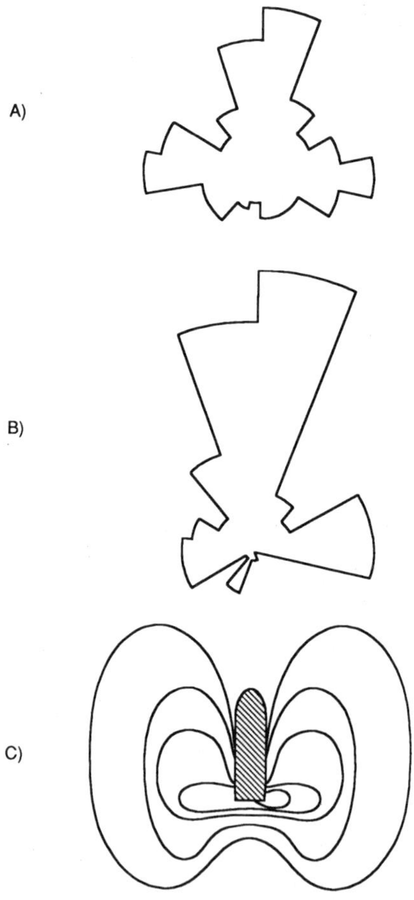

Figure 1-3. Distribution of radial swimming directions away from the survey vessel for schools of herring and sprat observed using a multibeam sonar: (A) 224 schools observed in front of a vessel at cruising speed; (B) 24 schools observed in front of the vessel during midwater trawling operations; (C) presumed noise directivity pattern for the vessel. After Misund and Aglen (1992).

alent when the vessel was engaged in midwater trawling (Fig. 1-3), due to the increased low-frequency noise associated with propeller cavitation.

Olsen et al. (1983) and Ona and Godø (1990) used stationary echo sounders to document strong avoidance reactions of herring (*Clupea harengus*), cod (*Gadus morhua*), polar cod (*Boreogadus saida*), capelin (*Mallotus villosus*), and haddock (*Melanogrammus aeglefinus*) as a survey vessel passed. Avoidance responses were observed for haddock aggregations as deep as 200 m. These reactions occurred primarily in response to noise produced by the survey vessel and involved both vertical (Fig. 1-4) and horizontal movements of fish. While vertical movement toward the bottom could actually enhance trawl catches, horizontal movements could place fish well outside of the area swept by the trawl doors, and reduce catches. The intensity of the avoidance response of haddock varied with propeller noise, and was observed to occur as much as 300 m in advance of the survey vessel when trawling was under way and the propeller was cavitating (Ona and Godø 1990). However, Atlantic herring at depths of 25–110 m and polar cod at depths of 110–220 m have been observed to initiate avoidance responses 75–150 m ahead of a survey vessel running at 10–12 knots without any fishing gear deployed (Olsen et al. 1983).

Fish are capable of directional hearing and can discriminate between sounds that are separated by 20° azimuth on the horizontal plane (Popper and Coombs 1980). Cod and probably many other species of fish are well able to locate sound sources in three dimensions, and can even determine the distance to the sound source (Hawkins 1986).

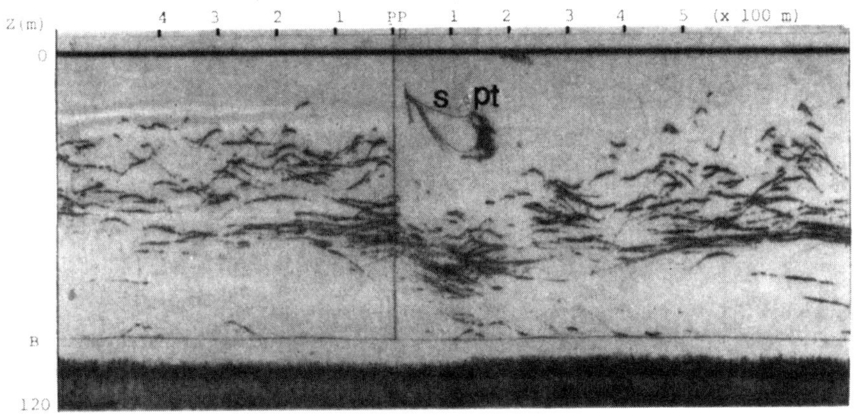

Figure 1-4. Vertical avoidance of a survey vessel by haddock, as observed from a stationary launch. Distance (meters) from the launch to the survey vessel is shown at the top of the figure, while depth (*Z*) is shown on the left. Note the sudden response to propeller noise (PP). The sweeplines (s) and pelagic trawl (pt) are visible, and the bottom (B) is indicated by the trace above the "whiteline." From Ona and Godø (1990).

Auditory capabilities vary widely between species, and are augmented by the presence of a swimbladder and coupling systems between the swimbladder and inner ear. The hearing abilities of most fish are confined to low frequencies, however (Fig. 1-5), and they are relatively insensitive to frequencies above 2 or 3 kHz. Atlantic cod have been shown to be sensitive to sounds in the 0.1–10 Hz frequency range (Sand and Karlsen 1986).

A wide variety of sounds in the frequency range of 0.03 to 3 kHz are produced by the engines, hydraulic winches, and propellers of survey vessels (Schwarz and Greer 1984; Mitson 1989). The larger the ship, and the faster it travels, the greater the distance at which it can influence fish behavior. However, even 7-m gillnetters and 25-hp skiffs have been observed to elicit an avoidance response in Pacific herring (Schwarz and Greer 1984), and avoidance of a rubber dinghy powered by a 20-hp outboard motor has been documented for 7–20 cm plaice (*Pleuronectes platessa*) in the Wadden Sea (Kuipers 1975).

Ideally, the noise produced by a survey vessel (including manned submersibles and ROVs) should be measured over an appropriate range of operating conditions, i.e., trawling, running speed (for acoustic surveys), etc. Remedial action can often be taken to change those characteristics of the

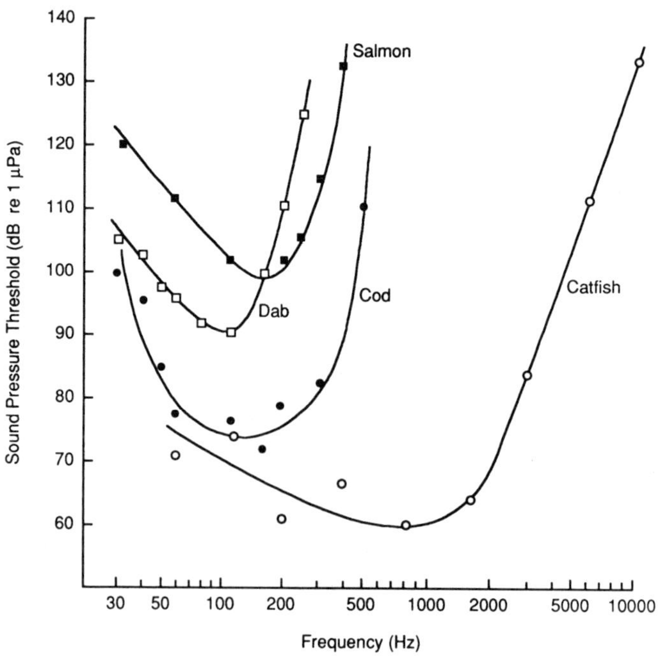

Figure 1-5. Audiograms for salmon (*Salmo salar*), dab (*Limanda limanda*), cod (*Gadus morhua*), and catfish (*Ictalurus nebulosus*) showing minimum sound levels detectable at a range of frequencies. After Hawkins (1986).

vessel that generate significant noise over the range of frequencies detectable by the target species (Mitson 1989).

Gear Avoidance and Selectivity

Once in contact with the sampling gear, fish can escape detection by either actively avoiding it or by failing to be captured once they enter it. It is consequently useful to define two different attributes of the target species: vulnerability and selectivity. Vulnerability can be defined as the probability of entering the gear (or area insonified in the case of acoustic samplers) given that an individual is in the path of that gear. In the case of beam trawls or plankton samplers with rigid frames, for example, the mouth area is well defined and any individual in the path of the gear will either elude capture by taking evasive action or end up inside the net. Only the latter are really vulnerable to the sampling gear; the former are virtually invisible by virtue of their ability to detect either the sampling gear or vessel deploying it and successfully evade it. Vulnerability, then, is a function of the sensory and locomotor skills of the target species, and, for any given species, these usually increase with individual size (Wardle 1983; Miller et al. 1988). Environmental conditions, notably ambient light and temperature, influence vulnerability (Wardle 1983; Glass and Wardle 1989; He 1991) as do the noise, turbulence, and light fields generated by the sampling gear (Murphy and Clutter 1972; Wardle 1983) and the vessel deploying it (Ona 1989).

Selectivity can be defined as the probability that an individual will be retained by the sampling gear given that it is vulnerable. Selection can be a relatively passive process, individual fish or eggs filtering through the mesh openings of the net (or failing to be detected by an acoustic beam) by virtue of their small size.

For most species and gear types, selectivity increases with size, while vulnerability decreases. The size-retention curve (Fig. 1-6) can be expected to be bell shaped or double-logistic, with the ascending limb being made up of those individuals that are large enough to be selected, and the descending limb comprised of those fish that are not mobile enough to avoid the sampling gear. The exact shape of this curve can be determined only when the size composition and abundance of the target population are known in detail. This is surprisingly rare, given the importance of this process in fisheries management, although the results presented by Myhre (1969, Fig. 1-7) for Pacific halibut offer one exception.

The selection processes that dominate the ascending limb can be readily quantified by comparing catches made with different mesh sizes in the case of trawls and plankton samplers, or by target strength studies with acoustic sampling systems. To evaluate avoidance of the vessel or sampling gear, however, either the abundance and size distribution of the target population must be known (preferably through tagging studies or stocking carried out prior to the survey), or direct observations must be made of behavior in the

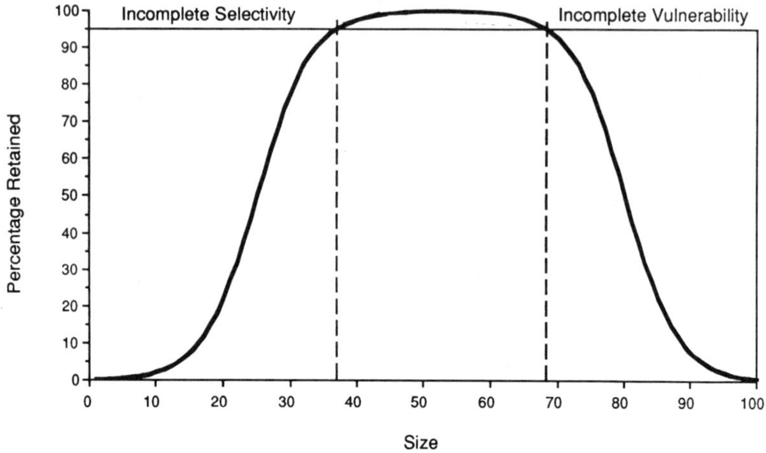

Figure 1-6. Relation between percentage retained and size, where the shape of the ascending limb is determined by selectivity, while the descending limb is determined by gear avoidance.

path of the sampling gear (Harden-Jones et al. 1977; Olsen et al. 1983). A common approach among ichthyoplankton researchers has been to estimate vulnerability through comparisons of day versus night catches (Morse 1989) on the assumption that no net avoidance occurs at night. These studies may be compromised where sensory abilities other than vision are important in the avoidance response, or where the turbulence generated by the sampling gear generates bioluminescence (Glass et al. 1986).

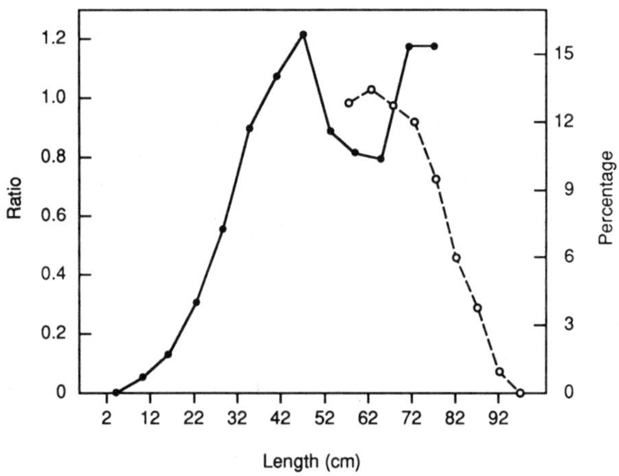

Figure 1-7. Gear selectivity and avoidance in Pacific halibut, based on the ratio of catches from a trawl with an 8.9-cm codend mesh with those of a control in parallel fishing experiments (solid line), and percentage of tagged fish recaptured (dashed line). From Myhre (1969).

Adequate quantification of avoidance processes is consequently the exception rather than the rule in fisheries surveys, a practice that has often resulted in wasted time and resources, and occasionally led to considerable embarrassment on the part of managers. It is often possible to make certain assumptions about the nature of avoidance and then attempt to validate them indirectly by comparing survey abundance estimates with results from independent methods (e.g., tagging studies or catch-at-age analysis). However, uncertainties surrounding the assumptions that underlie each estimate usually allow the investigator to tweak each estimate until the two agree, and can turn this process into a tennis match with no net. There is really no substitute for direct studies of avoidance and validation of vulnerability assumptions, and they should be an integral part of any direct survey program.

STATISTICAL ASPECTS

Bias

The final assumption in our quadrat sampling model is that the sites selected for sampling are a representative selection from the survey area, and that the mean abundance observed in these sites can be extrapolated to the entire survey area. This means that the sampling sites must be selected in such a way that the estimate is unbiased (Fig. 1-8). Of course, biased estimates will result whenever any of our assumptions about the area being sampled by the gear, or the availability, vulnerability, or selectivity processes operating during the survey are invalid. However, bias can also result from the manner in which the sampling sites are selected, although statisticians are divided as to how this is best avoided (Hurlbert 1984).

Strictly random selection of sampling stations is one possibility. In this instance, the entire survey area is subdivided into numbered sampling stations, and a random number generator used to select those stations that will actually be sampled. In practice, however, the pattern of stations resulting from such a random sampling procedure can be somewhat uneven. Several stations may be clumped in a small portion of the survey area, while other areas are covered only lightly. One way to minimize this is to constrain the randomized selection procedure so that no two sampling stations can be adjacent, although this represents a departure from the assumptions that underlie the computation of variance estimates. Nevertheless it is seldom possible to sample all of the sites selected in a strictly random design owing to unfavorable weather, clogging of nets, untrawlable bottom, mechanical breakdowns, and other chance events. For these reasons, strictly random sampling is an unrealistic goal in field surveys, although the data generated from them are routinely treated as though they come from a random sample.

Since gradients in density often occur with depth, latitude, salinity etc., a systematic selection of sampling stations (tracklines, grids, etc., Fig. 1-9) can

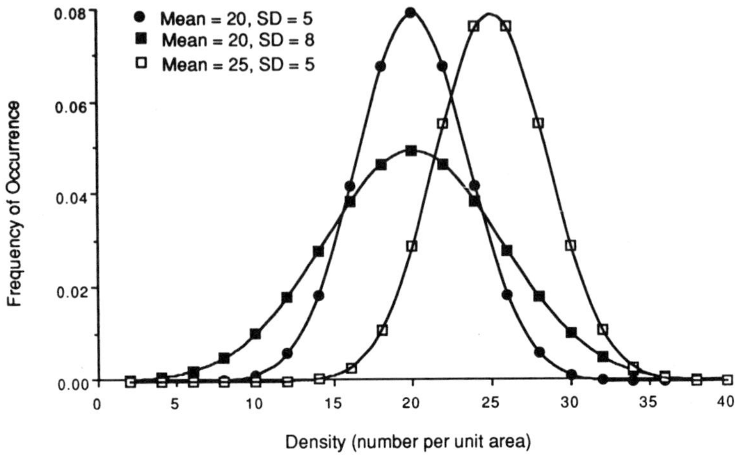

Figure 1-8. Bias and precision in surveys when sampling a population with a known density of 20 individuals per unit area. Circles represent density observations made using unbiased ($\mu = 20, \bar{x} = 20$) gear and sampling design, and characterized by high precision (SD = 5). Solid squares represent density observations made using unbiased gear and survey design, but with low precision (SD = 8). Open squares represent density observations made using biased sampling gear or survey design ($\mu = 20, \bar{x} = 25$), but still characterized by high precision.

often allow a more adequate dispersion of sampling stations than a random design, and minimize bias. The variance of a systematic sample can only be approximated from the results of a single survey, however, and there is a tradeoff between minimization of bias and exact estimation of variance in such instances. The logistics involved in sampling along equally spaced track-lines or gridlines often result in considerable savings in running time between sampling stations, however, and frequently tilt the balance in favor of systematic surveys. California Current ichthyoplankton surveys and Bering Sea trawl surveys routinely conducted by the U.S. National Marine Fisheries Service (Figs. 4-6 and 2-17) are good examples of systematic survey designs, and most acoustic surveys consist of observations made along systematic transects.

Precision

Constraints Imposed by Spatial Distribution

While the elimination of bias is the most important objective in survey design, high precision (Fig. 1-8) is also important. To a large extent, the maximum precision attainable is determined by the spatial distribution of the target species, surveys of patchily distributed species resulting in the lowest precision. Clumping, aggregation, or contagion characterizes most populations in the wild, and several models have been used to describe the frequency distributions of catches or density observations that result during a survey. A number of field studies (Lenarz and Adams 1980; Jahn 1987) have demon-

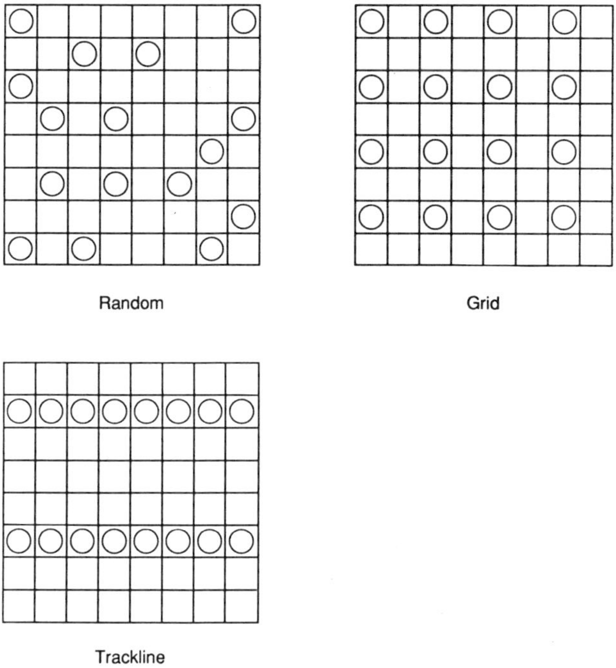

Figure 1-9. Alternatives in selecting 16 sampling stations from a total population of 64. Only one of these designs is based on a random selection of stations, while the others represent systematic grid (every other station in each row and column) and systematic trackline (all stations in every fourth row) designs.

strated that the negative binomial distribution is usually adequate to describe these, however. The variance (σ^2) of the catches (or density observations) increases with the mean in the negative binomial distribution, according to the relation:

$$\sigma^2 = \mu + \mu^2/k$$

where μ = arithmetic mean of the catches or density observations and
k = a parameter inversely related to the clumping of the population.

This model is sufficiently flexible to fit a wide variety of catch–frequency distributions (Elliott 1971).

The variance of the catches increases with their mean and degree of clumping when they follow a negative binomial distribution. The abundance and degree of clumping that characterize a population obviously play a major role in determining the precision that is ultimately attainable for a given survey, and the sample sizes required to achieve useful results. Estimation of the parameters of the negative binomial distribution is not a common practice,

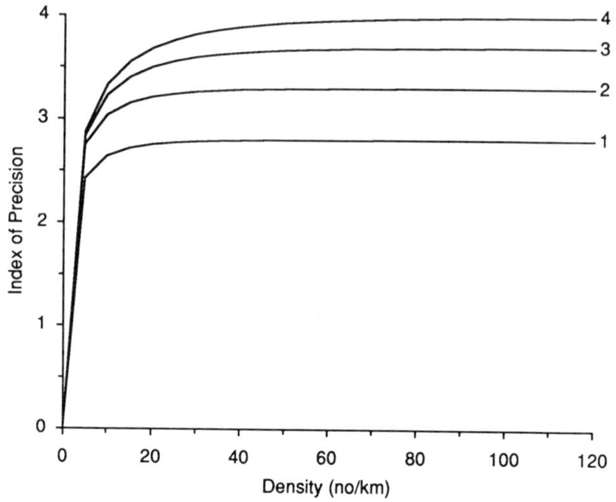

Figure 1-10. Alternatives in allocating the work day (roughly 13 hours) during trawl surveys of yellowtail rockfish (*Sebastes flavidus*). Strategies (bottom to top) are (1) three hauls/day with gear at depth for 2.1 hours, (2) four hauls/day with gear at depth for 1.2 hours, (3) five hauls/day with gear at depth for 0.5 hours, and (4) six hauls /day with gear at depth for 0.3 hours. After Lenarz and Adams (1980).

but in those cases where this is done, considerable insight is gained into sample sizes required. Tradeoffs between sample element size (e.g., quadrat size) and number of sampling sites to be occupied also become clear (Fig. 1-10).

Estimation of Bias, Precision, and Confidence Intervals

The relation between bias, precision (variance) and accuracy is shown graphically in Figure 1-11. The accuracy of any given density observation (x_i) is a measure of its overall deviation from the true value for the population (μ). Accuracy is also referred to as mean square error or MSE, and is the resultant of both bias and variance. $\text{MSE} = E(x_i - \mu)^2 = \text{bias}^2 + \text{variance}$.

Note, however, that bias cannot be estimated from the data collected during a survey. To estimate bias, an exact estimate of the true density of the target population would be required, an impossibility in most field surveys. As a result, the only way that bias can be evaluated is by comparing estimates of density or population abundance obtained by using independent techniques. A common way to approach this is to compare population estimates obtained through catch-at-age analysis with those from a direct survey (or surveys). Uncertainties exist in the key parameters used to generate any of these population estimates, however, and the process of "tuning" two potentially biased estimation techniques so that they eventually agree can be highly subjective. For this reason, independent validation of key assumptions (age determina-

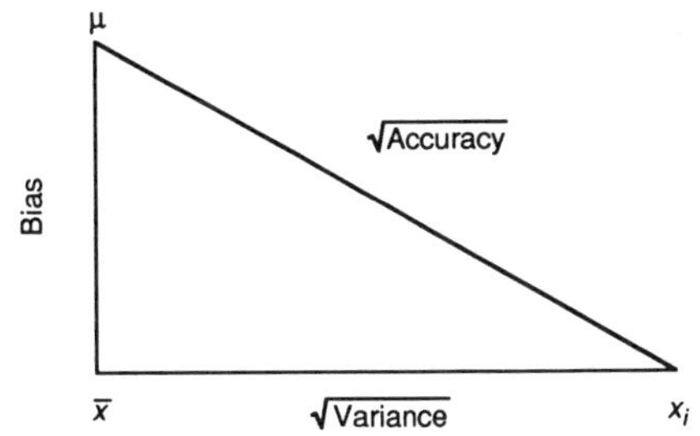

Figure 1-11. Relation between bias, accuracy, and variance.

tion, natural mortality, vulnerability, acoustic target strength, etc.) should be an important part of any stock assessment program.

Unlike bias, variance can be estimated directly using the data collected during the survey. Variance is simply a measure of the scatter of density observations around their mean, and can be readily quantified to make statements about the theoretical probability distribution of means that would be obtained if an infinite number of surveys were conducted.

Variance Estimation. Assuming that the survey area has been divided into strata, variance estimation is most straightforward if the sampling sites within each stratum have been selected randomly. With this design, each stratum is subdivided into non-overlapping units, and stations are selected by using a random number generator or random number table. An abundance estimate ($\hat{P}$ = population number, $\hat{B}$ = population biomass) can then be obtained from the catches (number or weight) at each sampling site. The estimate of P, for example, is

$$\hat{P} = \sum_{i=1}^{h} \left(\frac{A_i}{a}\right) \overline{C}_i \qquad (1.1)$$

where $\hat{P}$ = total population in survey area
A_i = area within stratum i (m^2)
a = area sampled by a single sampling unit
$\overline{C}_i$ = mean number of individuals caught per sampling unit in stratum i, based on n_i samples = $1/n_i \sum_{j=1}^{n_i} C_{ij}$
C_{ij} = number of individuals caught in the jth sample taken in stratum i
h = number of strata.

The variance of the population estimate is simply the sum of the within-stratum variances and is given by

$$\text{Var}(\hat{P}) = \sum_{i=1}^{h} \left(\frac{A_i}{a}\right)^2 \text{Var}(\overline{C}_i) \qquad (1.2)$$

where

$$\mathrm{Var}(\overline{C}_i) = \frac{1}{n_i(n_i - 1)} \sum_{j=1}^{n_i} (C_{ij} - \overline{C}_i)^2$$
$$= s_i^2/n_i$$

where s_i = standard deviation of the C_{ij} within stratum i.

In many instances, the area sampled (a) is difficult to hold constant (trawl hauls, icthyoplankton tows, etc.), and varies between sampling units. In this situation, all catches or counts can be normalized to a unit area (e.g., number/m^2 or weight/m^2) and become observations of density (D). This can be treated as a special case where $a = 1$, and D substituted for C throughout the development of Equations 1.1 and 1.2.

Where the total area sampled represents a large part of the total survey area, it may be desirable to use a variance estimator that incorporates the finite population correction $1 - n/N$ (Schaeffer et al. 1986) where N = total number of sample elements in the survey area, and n = the number of these actually sampled. For most survey applications, however, only a very small part of the total survey area is actually sampled, and the finite population correction is close to unity.

As mentioned previously, the investigator may decide instead to set up a systematic grid or series of tracklines within each stratum (Fig. 1-9), either for logistic reasons (reductions in running time, ease in station selection, etc.) or to achieve a uniform interspersion of stations. Hydrographic or chemical data such as temperature, salinity, dissolved oxygen, nitrogen, and phosphorus are most useful when collected in this manner, and this may also influence the selection of a systematic design. Where analysis of community structure (recurrent group analysis, assemblage analysis, etc.) is to be carried out using the survey data, a systematic design is also best.

One problem with systematic designs, however, is that the true variance of the population estimate cannot be estimated using the results of a single survey. Repeated systematic surveys, with a different starting point selected randomly for each survey, would be required to do this. A separate $\hat{P}_k$ is then calculated for each of the n_s systematic samples, and a grand mean estimated from

$$\hat{P} = \sum_{k=1}^{n_s} \frac{\hat{P}_k}{n_s} \tag{1.3}$$

The variance can then be estimated from the deviations of the $\hat{P}_k$ around the grand mean:

$$\mathrm{Var}\ \hat{P} = \frac{1}{n_s(n_s - 1)} \sum_{k=1}^{n_s} (\hat{P}_k - \hat{P})^2 \tag{1.4}$$

Schaeffer et al. (1986) suggest that n_s should be 10 or greater to obtain a satisfactory estimate of Var $\hat{P}$, however, so that repeated systematic sampling is seldom practical in field surveys.

Because this level of sampling is usually not feasible, the most common practice is to treat the samples obtained from a single systematic survey as if they were randomly selected, using Equation 1.2 to estimate the variance. This yields unbiased variance estimates only if the target species itself is randomly distributed within the survey area. If the abundance of the target species varies cyclically within a stratum (e.g., along wind-generated foam lines that are equally spaced), use of Equation 1.2 will result in an underestimate of the true variance, while it would result in an overestimate where there is a cline or gradient in abundance (Schaeffer et al. 1986). Cycles in abundance can bias the population estimate as well as the variance estimate, and the degree of bias cannot be evaluated without special field or simulation studies carried out on a case-by-case basis. Although it is difficult to generalize from the few studies that have been carried out, they have shown that systematic designs are more accurate and/or precise than random designs in the case of trawl surveys (Lenarz and Adams 1980), sonar surveys (Fiedler 1978), and phytoplankton surveys (Venrick 1978). Use of Equation 1.2 to estimate the variance of the population estimate from a single systematic survey would usually result in overestimation of the actual variance in the rockfish species studied by Lenarz and Adams (1980), resulting in confidence intervals that would be conservative.

Confidence Interval Estimation. If we can assume that the Central Limit Theorem holds, then the theoretical distribution of the population estimates would be normal, and the t-distribution can be used to construct confidence intervals around $\hat{P}$. With infinite degrees of freedom, for example, 95% confidence intervals will be $\hat{P} \pm 1.96\sqrt{\text{Var}(\hat{P})}$.

The number of observations required for the Central Limit Theorem to hold will depend on the distribution of the density observations. Aglen (1989) conducted replicate acoustic surveys in several different locations and found that survey estimates of mean density were normally distributed at higher levels of sampling intensity, but departed from normality when survey coverage was sparse. Simulation studies (Sissenwine 1978) have shown that the means from northwest Atlantic trawl surveys may not be normally distributed when sample sizes are less than 200.

Since violations of the normality assumption can result in erroneous confidence intervals, there has been considerable research into model-based estimation of confidence limits, where density observations are assumed to follow distributions such as the lognormal (Finney 1941) or the Delta distribution (Pennington 1983). However, model-based estimates of the population and its confidence interval can be biased and nonrobust to deviations from the underlying model (MacLennan and MacKenzie 1988; Smith 1990), and have seen limited use as a result.

Even when the distribution of the means is nonnormal, however, Tchebysheff's theorem can be used to construct conservative confidence intervals around $\hat{P}$. This theorem states that for any $k > 1$, at least $(1 - 1/k^2)$ measurements will lie within k standard deviations of their mean. Hence, at least $(1 - 1/k^2)$ or 75% of the population estimates generated from replicate surveys will lie within $2\sqrt{\mathrm{Var}(\hat{P})}$ (or 2 standard errors) of $\hat{P}$. As a result, the confidence interval $\hat{P} \pm 2\sqrt{\mathrm{Var}\ \hat{P}}$ will represent a 95% confidence interval if the Central Limit Theorem holds (replicate survey estimates are normally distributed) and a 75% confidence interval if it does not.

Maximizing Precision Through Survey Design

For any given spatial distribution of the target population, and any given level of survey effort, the precision attained can be increased substantially through proper survey design. This entails:

1. Effective stratification of the survey area into strata where density is homogeneous and intrastratum variance is low. Between-strata variance should be high, as this variance component does not enter into the estimate of the total variance for all strata combined. The variance of the population estimate for the total survey area is simply the sum of the individual within-stratum variances (Equation 1.2).

Strata are best defined on the basis of trends in the abundance of the target species, using data from previous surveys, or information on commercial catches. Clines in abundance typically follow well-defined bathymetric or geographic trends, however, so that it is usually possible to develop strata on the basis of depth, latitude, temperature or salinity even when data on actual abundance are unavailable. When the size and age composition of the population are to be estimated, stratification must maximize within-stratum homogeneity in these parameters as well as density.

A point of diminishing returns can be reached as the number of strata is increased. Using a regression model where $\phi(x) = E(y|x)$, Cochran (1977) examined the case where a variable x is used to stratify a data set y. If

$$y = \phi(x) + e$$

then

$$s_y^2 = s_{\phi(x)}^2 + s_e^2$$

Cochran's analysis suggested that $s_{\phi(x)}^2$ would decrease with the square of h (the number of strata), and that s_y^2 would approach s_e^2 with increasing h. At some point then, further increases in the number of strata will not reduce the between-strata component of variance. Based on linear regression models and consideration of three data sets available, Cochran concluded that little reduction in variance can be expected when the number of strata exceeds 6.

An analysis of trawl survey data for cod off Nova Scotia (Gavaris and Smith 1987) supports Cochran's conclusion.

A special case exists when postsurvey inspection of the data shows clear boundaries beyond which all catches were zero. This often occurs in the case of acoustic or ichthyoplankton surveys, where surveys are adaptive, and intentionally extended beyond the point where the target species actually exists. In the case of ichthyoplankton surveys, for example, stations are sometimes extended along a given transect line until two adjacent stations fail to detect any eggs. Since the spawning area can change substantially over time, this enables the investigators to map the area of occurrence during each survey.

In this situation, little bias is incurred by poststratifying the data into one stratum containing stations beyond the geographic limits of the target species (all catches are zero), and a second stratum containing all stations with nonzero catches. Several zero catches will still be embedded within the second stratum. This reduces the overall impact of nonzero catches on the variance, at a relatively small (if any) cost in terms of biasing the estimate of variance. The adjusted variance of such poststratified data has been estimated as $(1 + 1/n)\ \mathrm{Var}(\hat{P})$, where n = total number of stations, including those in the area where all catches were zero (Picquelle and Stauffer 1985). However, this is arguably not a situation where poststratification in the formal sense has actually occurred, and some authors simply ignore the $(1 + 1/n)$ correction term (Armstrong et al. 1988).

2. Effective allocation of survey effort between strata. The allocation that maximizes the precision of the population estimate for the total survey area (optimal allocation) will be such that the proportion of the samples in each stratum (n_i/n) is given by

$$n_i/n = \frac{A_i s_i/\sqrt{c_i}}{\sum_{i=1}^{h} A_i s_i/\sqrt{c_i}}$$

where A_i = area (m^2) of stratum i
s_i = standard deviation within stratum i
c_i = cost in money or time associated with obtaining a sample in stratum i
h = number of strata.

Estimates of within-stratum variance usually are not available, unless a preliminary survey has been conducted with allocation in mind. Francis (1984) recommends a two-phase approach to surveys, with the information available from an initial pass through the survey area being used to allocate effort during a second pass. As a rule of thumb, Francis suggests using about 75% of total survey effort during the first phase and 25% during the second.

Movements of fish between strata can occur even when a two-phase design is used to allocate sampling effort, however, so allocation based on variance estimates from previous years cannot be expected to be optimal. Nevertheless, some gains in precision can be expected even when only rough estimates of within-stratum variance are available. As a first approximation, we can allocate on the assumption that catches or density observations will follow the negative binomial distribution, and that the variance in each stratum is proportional to the square of the density there. This assumption is supported by empirical results from trawl (Fig. 1-12) and acoustic (Jolly and Hampton 1990) surveys. This approach has the added advantage of placing the greatest proportion of survey effort in those strata where the target population is most abundant, a feature that is intuitively appealing to any manager using the survey results, or user group whose livelihood depends on the survey. Fishermen put considerable effort into searching out areas of highest abundance, and find little comfort in being assured that while a survey may have missed many of these areas, station selection was strictly random, and the results unbiased.

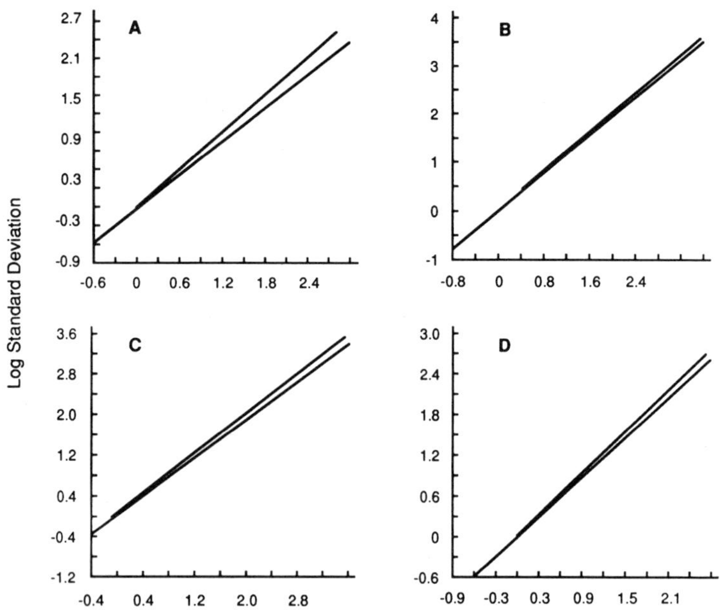

Figure 1-12. Regressions of logarithm of individual stratum standard deviation (Y-axis) against logarithm of mean catch per tow, for trawl surveys of the Gulf of St. Lawrence 1970–1979. The upper line in each case is a reference line with a slope of 1.0. Means of zero were omitted. (A) Gulf cod, (B) Shelf cod, (C) Gulf plaice, (D) Shelf plaice. After Koeller (1981).

3. Selection of optimal size for the sampling unit (length of trawl haul, size of plankton net or quadrat sampler, etc.). Since financial and manpower resources available to conduct a survey are always limited, tradeoffs exist between the size of the sampling unit and the number of samples that can be taken. A variety of studies have been undertaken to examine this tradeoff in trawl surveys (Lenarz and Adams 1980; Pennington and Grosslein 1978; Pennington and Vølstad 1991), surveys of aquatic vegetation (Pringle 1984), and surveys of benthic organisms with various quadrat samplers, corers, and dredges (Downing 1979), and the universal conclusion is that the sampling unit should be as small as feasibly possible. Shorter trawl hauls, and small-diameter samplers usually result in the highest precision attainable, since they allow the investigator to increase the number of these samples that are taken.

One example is the analysis of trawl surveys for rockfish (*Sebastes*) by Lenarz and Adams (1980). Using the negative binomial distribution, they developed an index of precision (P_d) that allowed them to evaluate four different ways of structuring each survey day.

When n catches are distributed according to the negative binomial distribution, they will have mean $\overline{C}$ and

$$\text{Var}(\overline{C}) = \frac{1}{n}\left(\overline{C} + \frac{\overline{C}^2}{k}\right)$$

Since $\overline{C} = \overline{D}a$ where $\overline{D}$ = mean density (numbers per unit area) and a = area sampled per trawl haul

$$\text{Var}\overline{D} = \frac{1}{na^2}\left[\overline{D}a + \frac{(\overline{D}a)^2}{k}\right]$$
$$= \left[\frac{\overline{D}}{na} + \frac{(\overline{D})^2}{nk}\right]$$

An index of precision can be defined as

$$P_d = \frac{\overline{D}}{\sqrt{\text{Var}(\overline{D})}}, \text{ the inverse of the coefficient of variation for } D$$
$$= \frac{\overline{D}}{\left[\dfrac{\overline{D}}{na} + \dfrac{(\overline{D})^2}{nk}\right]^{1/2}}$$

Trawl catch data were used to estimate the parameters of the negative binomial distribution for three species of rockfish, and examine P_d as a function of sampling strategy. An average of 2 hours was allowed between hauls to allow for running time and time required to set the gear, retrieve it, and process the catch. Four possible sampling strategies were evaluated, each representing a different way of allocating effort during a work day of roughly 13 hours. The results (Fig. 1-10) show that the highest precision was obtained with the shortest trawl duration (six trawls per day with the gear at depth for 0.3 hour), but that the overall level of precision attainable was sharply constrained by the spatial distribution of the target species. Species with a patchy or clumped distribution (low k) had much lower P_d values than more uniformly distributed species, regardless of the sampling strategy employed.

There are also practical limits to how small the sample element size can be, however. In the case of trawl surveys, trawl hauls shorter than 15 minutes should be avoided due to difficulties in determining the time the gear is actually on bottom and fishing properly. Sinking time varies, as does the time it takes for the gear to leave bottom once retrieval has begun. Navigational errors add to the problems of determining the area sampled by the gear during a haul, and since all of these errors represent a relatively constant unit of distance, their relative error increases with shorter tows. The behavior of fish must also be considered, since sufficient time must be allowed for some target species to be herded into the trawl, tire and fall back into the cod end.

Approximate Sample Size Requirements

The approximate number of samples required within a sampling stratum can be determined by examining upper and lower limits to the variance of the catches. These variances can be obtained from previous survey results, although they will probably change from year to year.

An approximate 95% confidence interval around the population estimate within a stratum is given by

$$\hat{P}_i \pm 2\sqrt{\mathrm{Var}(\hat{P}_i)} \tag{1.5}$$

Dividing both sides of the equation by A_i/a, this becomes equivalent to $\overline{C}_i \pm 2\sqrt{\mathrm{Var}(\overline{C}_i)}$.

Substituting $2\sqrt{\mathrm{Var}(\overline{C}_i)} = 2s_i/\sqrt{n} = W$, and rearranging, we obtain

$$n = \left(\frac{2s_i}{W}\right)^2$$

where s_i = approximate standard deviation of density observations (C_{ij}) within stratum i (from previous surveys)
W = desired half-width of confidence interval
n = sample size required

For example, if it is known from previous surveys that $\overline{C} = 2.3$ animals per sample and $s_i^2 = 0.65$, and we desire approximate 95% confidence in-

intervals of $\pm 10\%$ ($W = 0.23$), then the desired sample size is 4(0.65)/0.053 = 49, and about 50 samples would be required.

Equations for determining the sample size required to attain a given level of precision can also be developed for stratified sampling schemes (Cochran 1977; Schaeffer et al. 1986) but require a knowledge of the variance within each stratum, and stipulation of the manner in which sampling effort will be allocated between strata. Sampling costs can be included in sample size calculations and optimum sampling strategies determined. These are seldom used in practice, however due to the fact that both within and between stratum variances can vary widely between years.

Schweigert et al. (1985, 1990) used an approach based on Cochran (1977) and Jessen (1978) in developing a three-stage (spawning beds, transects within spawning beds, quadrats within transects) design for demersal herring eggs. The goal was a sampling strategy that would provide acceptable levels of variance (standard error within 25% of the mean) for estimates of egg deposition by Pacific herring. The variances involved changed substantially between years and survey areas, so that extensive analysis was required.

A more reasonable alternative to setting a rigid variance requirement, then determining the sample sizes required is to calculate the precision attainable with a number of different sample sizes given appropriate estimates of variance. The goal here is to gain some idea of the sensitivity of precision to alternative sample sizes that may be used. A common finding is that above a certain threshold, increases in sample size result in only marginal increases in precision. For example, in the case of the herring egg surveys conducted by Schweigert et al. (1985), the coefficient of variation declined quite slowly with increases in transect density beyond 2–3 per kilometer (Fig. 1-13).

When survey design within strata departs from simple random selection of stations, however, determining the number of stations to be sampled becomes more difficult. This is particularly true when the sampling design consists of a fixed series of transect lines, with sampling stations being arrayed at fixed intervals or depths along each line. This design is a two-stage sampling design, with transect selection representing the first stage, and station selection the second.

When it is difficult to fit survey data into classical sampling theory, bootstrap methods (Efron 1982; Efron and Gong 1983) can be used. The bootstrap process essentially uses the computational power available with modern computers to conduct multiple simulated surveys, each simulation using a different random selection (with replacement) of sampling stations. These simulations can be made employing any sampling regime desired, and the means and variances from each series of simulations then compared.

Kimura and Balsiger (1985) took this approach in evaluating a survey design for a sablefish (*Anoplopoma fimbria*) abundance index, whereby a vessel traveled to selected transects each year, and occupied stations at preselected depths along each transect. In their application, each year's survey data were first arrayed so that they were indexed by transect and depth, and the following procedure employed:

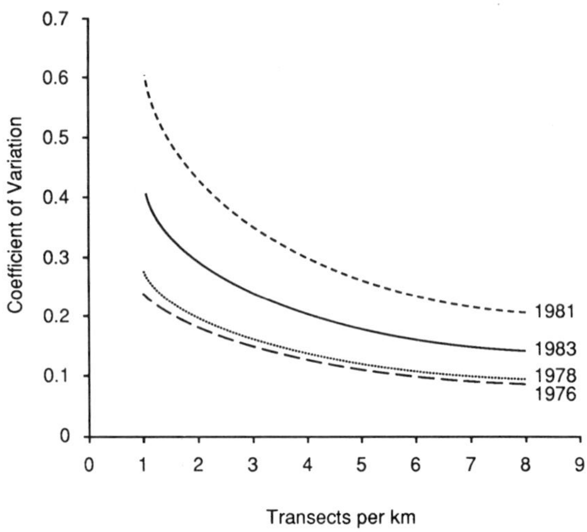

Figure 1-13. Relationship between the coefficient of variation (CV) and number of transects per kilometer for surveys of demersal egg deposition by British Columbia herring. After Schweigert et al. (1985).

1. A transect was randomly selected.
2. At each depth within the selected transect, a specified number of stations were selected and the "observed" catch per station recorded.
3. Steps (1) and (2) were repeated until the required number of transects had been sampled. Using the bootstrap, sampling is always with replacement.
4. The mean catch ($\overline{C}_i^b$) was calculated for each simulated survey.
5. The entire process (Steps 1–4) was repeated 1000 times, each iteration representing a simulated survey using a survey strategy of j transects and l stations per depth,
6. The mean and variance for each survey strategy was calculated from the individual ($\overline{C}_i^b$) values. These can then be interpreted as the mean and variance of a survey estimate when j transects are sampled each year and l stations are sampled per depth.

The results (Fig. 1-14) showed that the coefficient of variation ($CV = \overline{C}_i^b/\sqrt{\mathrm{Var}\overline{C}_i^b}$) was effectively reduced by increasing the number of transects, whereas increasing the number of stations per depth had little effect.

Transformations for Catch Data

When individual catch or density observations (as opposed to their mean) are to be used in statistical analyses that assume normality (e.g., analysis of variance, linear regression, or t tests), data transformations are often required.

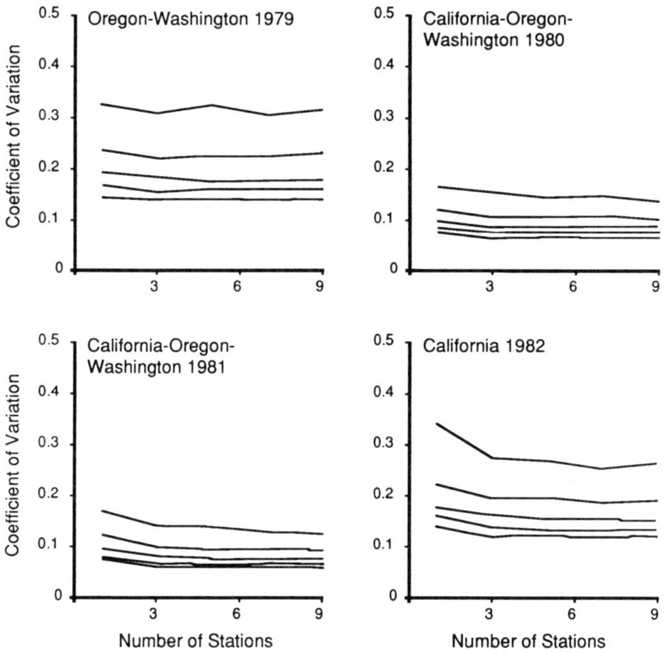

Figure 1-14. Bootstrap estimates of the coefficient of variation for pot surveys of sablefish off the U.S. coast. On each graph, curves (top to bottom) are for 3, 6, 9,12, and 15 transects sampled. After Kimura and Balsiger (1985).

If a transformation is effective, the transformed catch data will follow a normal distribution and their variance will be a constant, independent of their mean.

Two commonly recommended transformations (Taylor 1961; Elliott 1971; Sissenwine 1978; Downing 1979) are the log transform:

$$C_{ij} = \log(C_{ij} + 1)$$

and Taylor's power law transform:

$$C_{ij} = (C_{ij})^{1-b/2}$$

where b is derived from the relationship between log variance and log mean:

$$\log \mathrm{Var}(C_{ij}) = \log a + b \log \overline{C}_i$$

These transformations achieve homoscedasticity (variance homogeneity) when $\mathrm{Var}(C_{ij}) \approx a\,(\overline{C}_i^2)$ and $\mathrm{Var}(C_{ij}) \approx a\,(\overline{C}_i^b)$, respectively.

Using bottom trawl data for fish, Sissenwine (1978) considered the use of both transformations, and found that the resulting frequency distributions were similar in shape. Sissenwine concluded that there was no benefit in using the less commonly employed power law transform. Downing (1979) also explored the effectiveness of both the log transform and the power law transform, using density observations for a variety of benthic invertebrates. In this case, the power law transform proved superior by virtue of the fact that it always stabilized the variance, whereas the variance was occasionally correlated with the mean when the log transform was employed.

It should be remembered that a given transformation may not be effective, either failing to normalize the data (Pennington and Grosslein 1978) or to stabilize the variance (Downing 1979). Statistical tests should always be carried out on the transformed data to be certain that they fulfill all assumptions required.

2

BOTTOM TRAWL SURVEYS

TRAWL SYSTEMS

Vessel

Regardless of the type of trawl gear employed, the vessel selected for the survey will have an important influence on the overall effectiveness of the survey in achieving its desired results. The quality of the winches, lifting gear, navigation, and positioning equipment must all be high if the survey is to proceed rapidly and with accuracy. The experience of the captain and crew is also a major factor in determining the effectiveness of the vessel. Experienced fishermen are able to make informed decisions as to whether a given area is trawlable, spot and correct poor gear performance, repair damaged gear, and make sure that proper safety standards are followed throughout the survey.

To ensure smooth communication between the captain and chief scientist, a meeting should be arranged well before the survey begins. The captain should be provided with an overview of survey objectives, a copy of the trawl stations to be sampled, an itinerary for the survey, and copies of schematic diagrams that show the design and standard rigging of the trawls that will be employed. Procedures for deploying the gear, including the amount of trawl warp to be let out at any given depth, time allowed for the gear to sink, towing speed, and methods used to determine and record vessel positions at the start and completion of each tow should also be established at this meeting.

Otter Trawls

Otter trawls are the most widely used type of bottom trawl in commercial fisheries, and owe most of their success to the herding action of the trawl

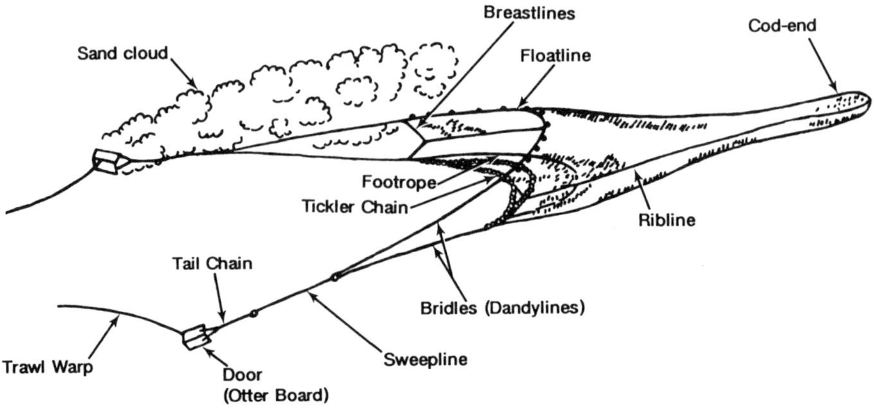

Figure 2-1. Schematic drawing of an otter trawl in action, showing the herding effect of the sand (or mud) cloud generated by the trawl door.

doors and sweeplines (Fig. 2-1) used. A substantial body of research on the behavior of fish in the path of otter trawls (Main and Sangster 1981, Wardle 1986) has shown that the trawl doors, sweeplines, and mud or sand clouds they produce present strong auditory and visual stimuli that can actually herd fish or shellfish into the path of the oncoming net. To be effective, however, the length of the sweeplines, attack angle of the sweeplines, and vessel speed must be carefully "tuned" to the maximum swimming speed of the target species. It is also important that the sand cloud generated be sufficiently high to herd the target species, and pass through the wing-ends of the net.

For a fish responding to the trawl door or sweepline to be captured by the net, the maximum value that the attack angle of the sweeplines can attain $(\theta) = \sin^{-1}(V_s/V_t)$, where V_t = trawl towing speed and V_s = maximum swimming speed of the target species. This follows from the need for the fish and the net to arrive at the same location (C, Fig. 2-2) at the same time, and the two equations

$$T_t = AB/V_t \cos \theta$$

and

$$T_s = AB \tan \theta / V_s$$

where T_t and T_s are times for arrival of the trawl and target species, respectively (Newland and Chapman 1989). Since V_s varies directly with fish length and ambient temperature (Wardle 1977, Fig. 2-3; He 1991), it follows that the attack angle of the sweeplines cannot be optimized for all sizes of fish, and the herding action of the sweeplines will be highly size selective. Experimental work by Engås and Godø (1989a) has borne out theory in this case,

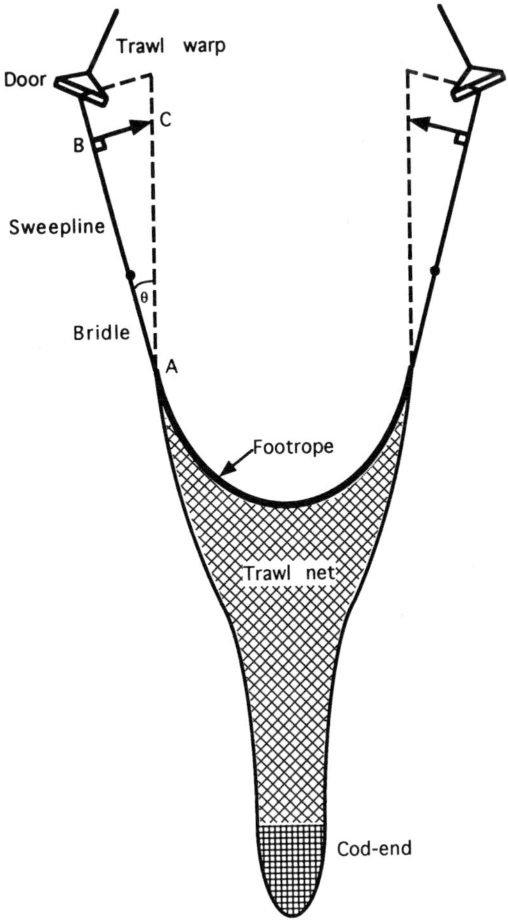

Figure 2-2. Trajectory (BC) followed by a fish avoiding the trawl doors, sweeplines, and bridles, and ending up in the net (A). θ = attack angle of sweeplines.

and shown that changes in sweepline length can drastically alter both the magnitude and size composition of cod catches.

The situation is complicated further by the behavior of the target species in response to the doors and sweeplines, and variability in both the attack angle of the sweeplines and the nature of the sand cloud when towing over different substrates. Direct observations have shown that while the auditory stimuli produced by the trawl doors increases when towing over hard bottom, visual stimuli produced by sand clouds are greatly reduced. Furthermore, spreading forces produced by the trawl doors are reduced, resulting in smaller distances between the doors and reduced attack angles for the sweeplines (Main and Sangster 1979).

A wide diversity in the behavioral responses of fish and shellfish to trawls has been observed directly using cameras or SCUBA divers (Main and Sangs-

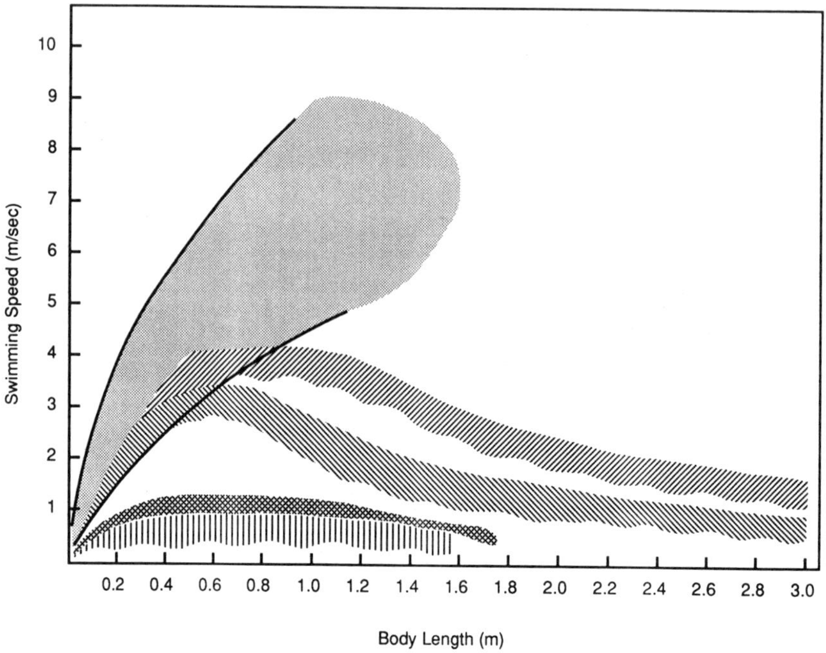

Figure 2-3. Relationship between swimming speed (m/sec) and body length (m). The cross-hatched areas represent estimated sustained swimming speeds of (top to bottom): Scombroidei, warm body in cold water; Scombroidei, warm body in warm water; Salmonidae and Gadidae. For comparison, the stippled area between the solid lines shows maximum burst swimming speeds for temperatures between 0 and 20°C and stride lengths between 0.6 and 0.8. After Wardle (1977).

ter 1981), and it is impossible to generalize these results. North Sea plaice (*Pleuronectes platessa*) tend to herd readily, moving perpendicular to the sweeplines in short bursts until they are in front of the footrope. They then swim in front of the net until they exhaust their anaerobic metabolic capabilities and fall back into the net. While some plaice escape under the sweeplines or footrope, proper selection of the mesh sizes used can ensure that few will escape once they are in the trawl itself.

Haddock form a column or ribbon of fish equidistant between the sweeplines, gradually losing ground relative to the approaching trawl. The typical behavior of a tiring haddock is to swim slowly upward and turn while rising from the sea bed. Some haddock rise high enough over the headline to escape. In contrast, cod tend to drop back close to the bottom upon tiring and many of them are able to turn and escape under the footrope. When conventional bobbin (roller) gear is used on the footrope, considerable numbers of cod, particularly the smaller (<35 cm) individuals, dive under the fishing line and escape (Engås and Godø 1989b).

Flatfish are also able to escape beneath the footrope. Walsh (1991) found that at least 95% of the juvenile (<23 cm) American plaice (*Hippoglossoides platessoides*) and yellowtail flounder (*Limanda ferruginea*) encountered by the survey trawl used off Newfoundland were able to escape beneath the bobbin-equipped ground gear, regardless of light conditions. Significant numbers of adults also escaped beneath the ground gear. Escapement was greater during the day (35–65%, depending on species and size) than at night (20–29%). Underwater video observations indicated that flatfish respond to the gear much sooner during the day than at night, and herding was recorded. Attaching a tickler chain to the ground gear (Fig. 2-1) increased catches of both juvenile and adult yellowtail flounder during the day (Walsh 1988).

Bobbin gear is essential when the target species inhabits areas where the bottom is rough (Gunderson and Sample 1980), and escapement under the footrope can be minimized by employing such new designs as "rockhopper" rigging of the bobbins (Engås and Godø 1989b, Fig. 2-4). The main innovation of rockhopper gear is the off-center attachment of the footrope to the rubber discs, which allows the footrope to tend bottom more consistently, and still protect the net. Rockhopper gear is also heavier than many bobbin gear configurations traditionally employed.

Reaction distances to visual stimuli are influenced strongly by ambient light conditions, and at extremely low light conditions ($<10^{-6}$ lux) herding response by haddock can be lost (Glass and Wardle 1989). However, studies of mixed aggregations of haddock, cod, and saithe (*Pollachius virens*) suggest that in some situations the fish were able to detect and respond to sound produced by the trawl gear and vessel towing it, even at night when light levels were about 2×10^{-6} lux (Engås and Ona 1990). It follows then that the area effectively swept will vary with depth, season, time of day, ambient light (including that provided by bioluminescence), and response to auditory stimuli produced by the survey vessel and gear. Schooling responses can either enhance (Ellis 1973) or reduce (Godø et al. 1990) the overall effectiveness of the doors and sweeplines, so that effective area swept can also vary with the density of the target species.

Reaction distances to the sweeplines are quite short for the Norway lobster (*Nephrops norvegicus*), and the herding response is still relatively complex.

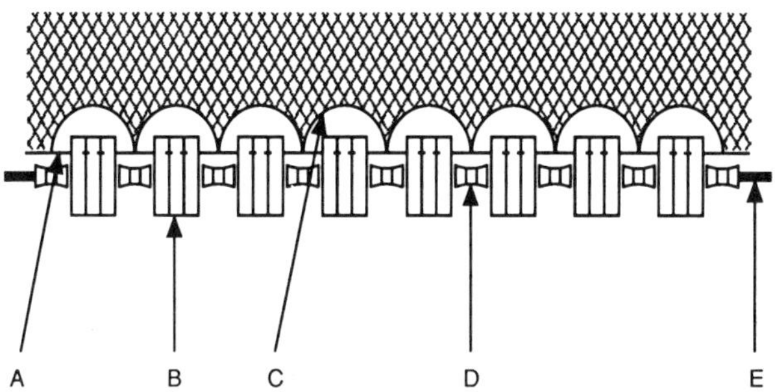

Figure 2-4. Schematic drawing of rockhopper ground gear: (A) footrope, (B) rubber disc (bobbin), (C) fishing line, (D) spacers, (E) chain. After Engås and Godø (1989b).

About half are touched by the gear before starting to swim (Newland and Chapman 1989). Those that face toward the gear react to it by swimming directly away at an angle of about 90°, while no herding response is shown if they face away from the gear. In the latter case, *Nephrops* respond by lifting their tails and swimming directly back over the sweeplines. Once near the trawl, *Nephrops* facing the gear must be induced to swim up and over the footrope to be captured, or they are likely to escape under the net.

Given all of these complications in the behavior of both otter trawls and target species, how can the effective area swept be determined? Research by Harden-Jones et al. (1977) provides considerable insight into the type of experiments required to address the problem. These investigators placed acoustic transponding tags on North Sea plaice, so that their exact position was known at all times. They then made 343 "attacks" on tagged fish using an otter trawl rigged so that the distance between the doors was 40 m, and the area covered by the path of the net was 18 m wide. The position of the fish being attacked, trawl doors, and net were monitored continuously by a second vessel equipped with a sector scanning sonar. Their results showed that 44% of the fish between the doors were captured, so that the effective area swept would be 0.44 (40 m) or 17.6 m. Only 61% of the fish that were in the path of the net itself were actually caught.

Such results pertain only to the gear, rigging, and towing speeds employed in their experiment, however, and cannot be generalized to other locations or even to plaice of a different size without additional research. Adequate definition of effective area swept by an otter trawl clearly becomes an expensive process if it is to be done properly. It is not surprising then that many researchers have resorted to the use of beam trawls (Fig. 2-5), where no otter boards are employed, for purposes of quantitative sampling.

Beam Trawls

Beam trawls are the oldest and simplest of trawl systems, and still enjoy widespread use in commercial fisheries. Dutch fishermen use large (12 m beam), heavy versions of the beam trawl in North Sea sole (*Solea solea*) fisheries, and usually tow two of these trawls simultaneously. These commercial trawls are quite heavy, and their rigid beam makes them difficult to store aboard small vessels and deploy. As a result most research applications employ relatively small (6 m beams or smaller, Fig. 2-5) versions of the beam trawl, particularly where it is necessary to sample in very shallow nearshore or estuarine habitats. As with otter trawls, "tickler" chain arrays are often used in front of the footrope to dig into the substrate and startle or dislodge organisms that will then be swept up into the water column and trapped by the net. Careful design of tickler chain arrays can eliminate most escape beneath the footrope, although it is important to carry out gear experiments to determine just how many strands of chain are required for a given substrate.

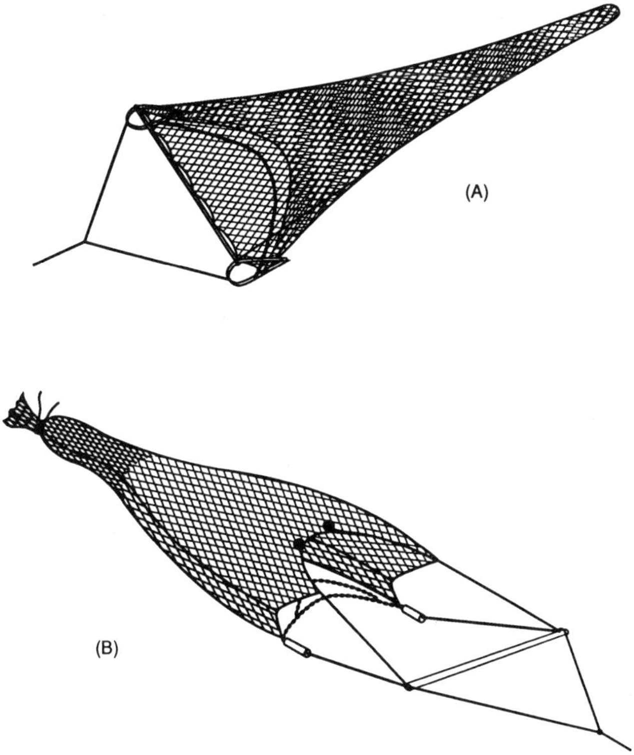

Figure 2-5. Beam trawl designs: (A) 2 m beam attached to "shoes" on each side. (B) Rigid frame replaced with removable aluminum staff and wingtip weights. Both nets are fished with tickler chains or tickler chain arrays in front of the footrope.

These experiments are usually quite empirical, increasing the number of tickler chains until catch rates become asymptotic (Kuipers 1975; Creutzberg et al.1987)

The principal advantage of the beam trawl in research applications is the well-defined area swept. Their main disadvantages lie in the heavy, cumbersome frames or beams traditionally employed in their construction, and the bridles used in their deployment. While recent designs have eliminated the former problem by replacing the traditional beam with one that can be detached and stored separately (Gunderson and Ellis 1986), the bridle problem remains. The bridles used with beam trawls probably frighten off many pelagic, mobile species in advance of the net. As such, its primary use should be in sampling fish and invertebrates that are demersal and show only limited escape responses. Juvenile flatfish and benthic invertebrates are excellent candidates for beam trawl sampling.

Monitoring Trawl Performance

A critical part of any quantitative trawl sampling program is the accurate determination of the distance fished. Instrumentation of each trawl haul should be the goal in any modern trawl survey, since research has shown that there is no other way to be certain whether the trawl is on bottom. The time a trawl takes to reach bottom after deployment varies significantly with prevailing currents, and can be difficult to predict. Godø and Engås (1989) found that in deep water the watch officer's estimate of the time of bottom contact was off by as much as 10 minutes. The time between initiation of the retrieval process and cessation of effective fishing also varies, and ranged from 0 to 5 minutes during 130 otter trawl hauls monitored by West (1985). More serious, however, are undetected failures in gear performance. Improper spreading of an otter trawl or failure for the net to be on bottom for substantial periods occurred on 6 of 26 (23%) of the Campel 1800 tows monitored by Engås and West (1987). Wathne (1977) found that 400-Mesh Eastern otter trawls were off bottom for substantial periods of time in 11 of 38 (29%) tows he monitored, while similar failures occurred in only 6 of 77 (8%) of the hauls he conducted with a modified (83/112) Eastern otter trawl. Both Wathne (1977) and West (1985) suggested that strong bottom currents were responsible for much of the erratic behavior they observed.

Measuring wheels and odometers mounted directly on beam trawls are one of the simplest monitoring devices, and are still used in some research applications. These devices are easily damaged on rocky bottom, however, and their performance can become erratic in some instances. A study conducted off the coast of Oregon indicated that odometer readings on a 3-m beam trawl underestimated the actual distance fished by about 40% (Carney and Carey 1980).

Acoustic monitoring systems that provide continuous information on the depth of the trawl, vertical and horizontal opening of the net, distance between trawl doors, and speed through the water (Fig. 2-6) are available commercially and enjoy widespread use in commercial fisheries where real-time information on trawl performance can increase both efficiency and earnings. Deployment of these same systems on full-scale trawls used routinely in quantitative resource surveys is strictly a matter of economics, and should become a routine practice wherever the management of important commercial fisheries is dependent on survey results. Research to date has shown convincingly that failure to monitor trawl performance properly will often introduce unacceptable levels of error into survey results.

Proper monitoring of smaller trawls currently presents a problem, however, since the systems developed for commercial trawls are too large to deploy on them. Compact time–depth recorders have been developed for research on marine mammal diving behavior, and are capable of storing data on trawl depth and downloading it to a computer after the haul has been completed. Acoustic monitoring systems that are developed specifically for use on small

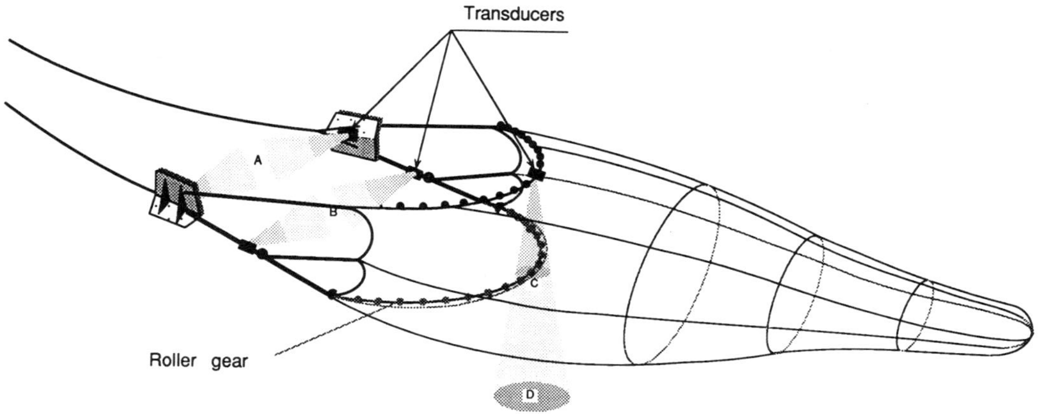

Figure 2-6. Acoustic monitoring of a trawl, using transducers arranged so as to enable mensuration of door spread (A), horizontal net opening (B), vertical net opening (C), and distance between footrope and bottom (D).

trawls can be expected to become available in the near future, however, so that real-time monitoring of trawl performance should become a routine matter during any quantitative trawl survey.

In those situations where budgetary considerations preclude instrumentation of the trawl and proper monitoring of its performance, the investigator should do as much as possible to minimize errors in estimating the distance fished. Vessel speed should be minimized during the deployment and retrieval of the trawl, to reduce the impact of errors in the estimates of the time the gear contacts and leaves bottom on the estimated distance fished. Trawling should be avoided in situations where bottom currents are strong, and could adversely impact gear performance.

Distance Fished

As discussed in Chapter 1, the distance fished should be kept as short as possible, owing to the patchy distribution of most species, and the resultant need to favor sample size over sample element size. Just how short should survey tows be? A number of experiments have been carried out in the Bering Sea, Georges Bank, and Barents Sea to address this question, using carefully designed experiments and analysis of variance techniques. The Bering Sea studies (Gunderson 1976a) consisted of two series of experiments where 25 sets of comparative hauls were conducted to examine differences between 30 and 60 minute hauls. Three vessels were employed, and made a total of 73 usable hauls. No significant differences were detected in the catch per square meter fished for pollock, tanner crab, or flathead sole, supporting the decision to adopt 30-minute hauls as the standard for Bering Sea surveys.

A series of 64 hauls were made at eight stations on Georges Bank (Godø et al. 1990), two hauls each of 15, 30, 60, and 120 minute duration. No significant differences in catch per minute or mean size of haddock were found between hauls of different duration. A series of 60 hauls in the Barents Sea yielded a similar result, no significant differences in catch per minute or mean size of cod, haddock, or long rough dab (*Hippoglossoides platessoides*) being detected between hauls of 15, 30, and 60 minute durations. However, a subsequent series of 24 hauls in the Barents Sea showed that significant differences in the catch per minute (but not mean size) of haddock and cod did exist between 5 and 30 minute hauls (Godø et al. 1990), with a higher mean catch per minute for the shorter tows. This suggested that some factor that enhanced "catching by surprise" rather than "catching by exhaustion" (e.g., reduced ability to respond as schools) actually makes shorter tows more effective than longer ones, so that tows as short as 5 minutes would be most desirable. In actual practice, a lower limit on tow duration is imposed by navigational error and the ability to determine exactly when the net contacts and leaves bottom, since these factors introduce an increasingly large percent error as tow length decreases.

Assuming that a similar amount of trawl warp is out when the gear first touches the bottom and when it leaves bottom, the distance fished can be determined quite accurately from the position of the vessel at these two points in time. Modern navigational positioning is accurate to within about 20 m when Loran C equipment is used to estimate the difference between two positions, and accuracy can be even higher with a satellite-based Global Positioning System. Variable range radar can also provide estimates that are accurate to within about 20 m in areas that are close to shore and where the topography is such that identifiable features are plentiful, although accuracy falls off sharply where conditions are less than optimal. Sextant fixes are affected by distance, visibility, and the presence of readily identifiable topographic features, but can be used when more reliable systems are unavailable.

When tows are short (less than about 600 m), reasonably accurate estimates of the distance fished can be obtained with relatively unsophisticated instruments. Marker buoys can be set at the points where the gear contacts and leaves the bottom, and the distance between them determined using an optical range finder. The accuracy of these range finders declines as a function of distance, but can be kept to within about 10% if sighting distance does not exceed 300 m. Sighting distance can be kept to a minimum by positioning the vessel midway between the two marker buoys, then sighting back at each of them to obtain the total distance fished.

Trawl Width and Door Spread

While the area swept by a beam trawl is fixed, and can readily be determined, otter trawl performance can be highly variable. Both the distance between the doors and the horizontal width (wingspread) of the trawl itself have been used by researchers as a measure of the effective area swept by an otter trawl. Research has shown that door spread and/or horizontal opening both vary significantly with bottom type (Main and Sangster 1979), scope (amount of trawl warp out, Rose and Walters 1990), depth (West 1982, Godø and Engås 1989), towing speed (Wathne 1977), bottom currents, and amount of fish in the cod-end (West 1985). As would be expected, the vertical net opening decreases with increases in wing spread (Godø and Engås 1989, Rose and Walters 1990), reducing the effectiveness of the trawl in capturing species (e.g., haddock) that attempt to escape over the trawl headrope.

As a result, it is desirable to instrument each trawl in such a way that door spread or horizontal width can be monitored continuously. Net monitoring systems that are capable of doing this have already been designed and field tested for commercial fisheries applications, and can be readily obtained. Rose and Walters (1990) have shown that biomass estimates for pollock could be biased by as much as 15% and the coefficient of variation increased as much as 5% by failing to monitor net width properly, and assuming it to be constant between surveys. Door spread is probably a more important measure of effective area swept than net width, however, and may vary independently.

Experiments by Engås and Godø (1989a) showed that increases in sweepline length resulted in increases in both door spreads and catches, even though the wing spread of the net changed little. Comparisons between door spread and wing spread where sweepline length was held constant and warp length varied (Koeller 1991) did show a significant increase in wing spread with door spread, but the slope of the regression equation ($b = 0.17$) was small, indicating a very flat response in wing spread to changes in door spread. Bias in the Scotian Shelf groundfish survey estimates of haddock abundance caused by unmonitored variations in effective fishing width would have been 15–27% if door spread were used to define fishing width, and 10–15% if wing spread were used.

In all comparisons made by Engås and Godø (1989a), the trawl with the longer sweeplines caught the most cod and haddock, and a higher proportion of large fish. Hence both catch and size composition can depend on door spread. At some point, however, increases in catch rate will become negligible or even decline with further increases in door spread (Bridger 1969, Harling and Davenport 1977). For species that are not herded by the sweeplines, net width is the main parameter determining catch, and the catch–door spread relation will rapidly become asymptotic as door spread is increased. For species that are herded, the catch–door spread relation will become parabolic as fish become unable to swim rapidly enough to reach the net, or the sand cloud generated by the doors fails to reach the net wings.

Since door spread can often influence both the magnitude and size composition of the catch, it is important that it be standardized. Engås and Ona (1991) have shown that strapping the trawl warps together in front of the doors is one way to restrict the variability in door spread. The restraining rope used to strap the warps together was slighty longer than the distance between the towing blocks, and was attached to the warps with slip hooks. In their application, an 8.9-m restraining rope attached 150 m in front of the trawl doors maintained a door spread of 51 m for warp lengths of 600–800 m.

An alternative was suggested by Koeller (1991) who systematically varied trawl warp at a several depths and found that door spread was highly correlated with the trawl warp/depth ratio (Fig. 2-7). Once these relationships have been worked out, it may be possible to use them to determine the trawl warp required to achieve constant door spreads at different depths. More important sources of uncertainty about the effective fishing width will still remain (i.e., those related to the sensory abilities and escape behavior of the target species), but a significant source of bias and variability will nevertheless be eliminated.

Estimating Gear Efficiency

Arguments can be made for using either the distance between the doors or horizontal net opening to the define the effective fishing width of an otter trawl. The parameter selected to define fishing width will depend on the target species and its herding behavior, and must be carefully standardized and monitored during the survey. Whichever measure is chosen as the effective fishing width, however, must be accompanied by estimates of the overall efficiency of the gear in catching individual fish that lie within this path. One

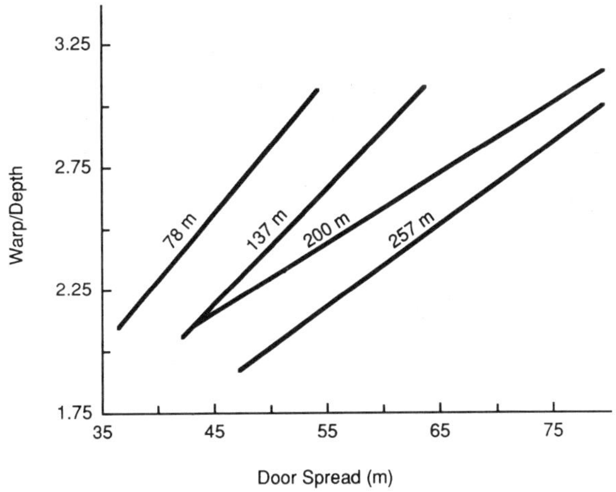

Figure 2-7. Relationship between door spread and warp/depth ratio at four different depths. After Koeller (1991).

problem in estimating gear efficiency is that many species react to the vessel towing the gear, and initiate avoidance responses well in advance of the oncoming trawl (Chapter 1).

Overall efficiency can be expected to vary widely with vessel characteristics, sensory capabilities, and behavioral responses of the target species, and a wide variety of environmental variables (light, temperature, bottom type, etc.). Efficiency will be a variable then, rather than a constant, and will clearly be difficult to estimate. For this reason, many investigators choose to treat density estimates from trawl surveys as relative indices rather than estimating abundance from the area swept. This in no way obviates the need to understand sources of bias or variability that influence the density estimate, however, and to correct for them wherever possible. "Relative" abundance indices have all too often been used to mask shortcomings in survey procedures, notably failure to adequately quantify the distance fished or area swept by the trawl.

Estimation of efficiency has been most successful where trawl surveys have been conducted concurrently with tagging or depletion (Leslie–DeLury) experiments. These experiments are often subject to biases of their own, but provide abundance estimates that are truly independent of the direct survey approach. Observations with submersibles and underwater cameras do not get away from basic problems with vessel avoidance, and the lights that must often be used with these devices result in further complications. For example, some species appear to be attracted to submersibles while others avoid them (Ralston et al. 1986, Pearcy et al. 1989).

Day–night comparisons of catch rates (Gunderson and Ellis 1986) can quantify the effects of visually cued avoidance, but sound-induced avoidance still goes on at night, and cannot be evaluated by such experiments. Furthermore, bioluminescence can provide sufficient light to make trawls and trawl gear visible at night, and can also confound day–night comparisons.

Defining the area swept as effective net width, Nielsen (1983) used Petersen mark–recapture experiments to estimate gear efficiency and its variance for yellow perch (*Perca flavescens*) in Oneida Lake, New York. This is a shallow (4–11 m), enclosed body of water, so that immigration and emigration were not a problem. Efficiency (E) was estimated as

$$E = \frac{A\overline{C}}{SP}$$

where A = total number of trawl units in the lake
$\overline{C}$ = mean catch per haul
S = survival rate between tagging and survey, and
P = population size at time of tagging, determined using the Petersen estimate

Variance of the efficiency estimate was derived from a Taylor-series expansion (or delta method, Seber 1973) as

$$\begin{aligned}\operatorname{Var}(E) = {} & \operatorname{Var}(\overline{C})\left[\frac{A}{SP}\right]^2 + \operatorname{Var}(P)\left[\frac{\overline{C}A}{SP^2}\right]^2 \\ & - \operatorname{Cov}(\overline{C}, P)\left[\frac{\overline{C}M^2}{P^3}\right]\end{aligned}$$

where M = number of fish marked

Nielsen was able to estimate efficiency by age group, but found no consistent differences for fish from age 3 to 7 years and older.

Efficiency for an otter trawl with a 5.9 m footrope ranged from 8 to 12% during 1970–1974. Large numbers of fish were marked each year (45,000–75,000) and at least 110 tows were made annually so that these efficiencies were determined with coefficients of variation that ranged from 27 to 55%. During 1969, however, water transparency was quite low and gear efficiency increased to 26% (CV = 12%). This suggests that gear efficiency could have been increased, and interannual variation reduced, if trawling were done at night (when visual cues would be minimized) rather than in the daytime.

Edwards and Steele (1968) tagged 300 to 1000 juvenile plaice in Loch Ewe, Scotland, held them for 24 hours to eliminate dead or weak individuals, then dispersed them evenly in a shallow (<8 m) bay about 350 m wide. Immigration and emigration were assumed to be negligible.

Approximately 36 hours later, a routine survey of the plaice population was conducted using a 2-m beam trawl. Efficiency was not estimated directly, but tagging results served to confirm the basic reliability of estimates determined independently by towing the beam trawl through an enclosure containing a known number of fish. These efficiency estimates ranged from about 25% for 20-mm fish to about 32% for 30- to 60-mm fish, and elucidated a design flaw in the gear that allowed the footrope of the trawl to lose contact on uneven bottom.

Leslie–DeLury depletion experiments have also been used to estimate population sizes and gear efficiency for species that are either sessile or capable of limited mobility. The survey area must be large enough that it can be assumed to be closed to migrations during the sampling period, however, limiting the applicability of these experiments for many species.

Iribarne et al. (1991) used a depletion experiment to estimate the efficiency of scallop dredges in the San Matias Gulf, Argentina. They set up a 366,606–m^2 study area, and carefully monitored the activity of commercial fishermen in the area, using onboard observers to obtain data on catch and catch per minute fished (CPUE). The data were then fit to both the Leslie and DeLury models using linear regression. These models were expressed as

$$\frac{C_t}{f_t} = qP_1 - q\left(K_t + \frac{C_t}{2}\right) \qquad \text{(the Leslie model)}$$

and

$$\ln\left(\frac{C_t}{f_t}\right) = \ln(qP_1) - q\left(E_t + \frac{f_t}{2}\right) \qquad \text{(the DeLury model)}$$

where C_t = catch during time t
f_t = effort during time t (in minutes fished)
q = coefficient of catchability
P_1 = initial population size
K_t = cumulative catch up to time t, and
E_t = cumulative fishing effort applied up to time t

Regression of C_t/f_t versus K_t or $\ln(C_t/f_t)$ versus E_t can then be used to obtain estimates of both q and P_1 (Fig. 2-8). Gear efficiency (E) can be defined as $(A/a)q$, where A and a are total survey area and area swept by a minute of fishing effort, and its estimation is relatively straightforward. Efficiency estimates ranged from 15 to 21%, depending on which model was used. The variance of E was obtained using a bootstrap procedure (Efron 1982), and the coefficients of variation for these estimates ranged from 36 to 42%. Observations by SCUBA divers showed that gear efficiency was being impaired by accumulation of a pile of material in front of the dredge and in the throat of the dredge during operation.

A somewhat different approach was taken by Joll and Penn (1990), who used a trisponder in conjunction with a high-resolution navigation system to maintain vessel position within a few meters. This allowed them to systematically trawl an entire 384,000 m^2 survey area on three successive nights of

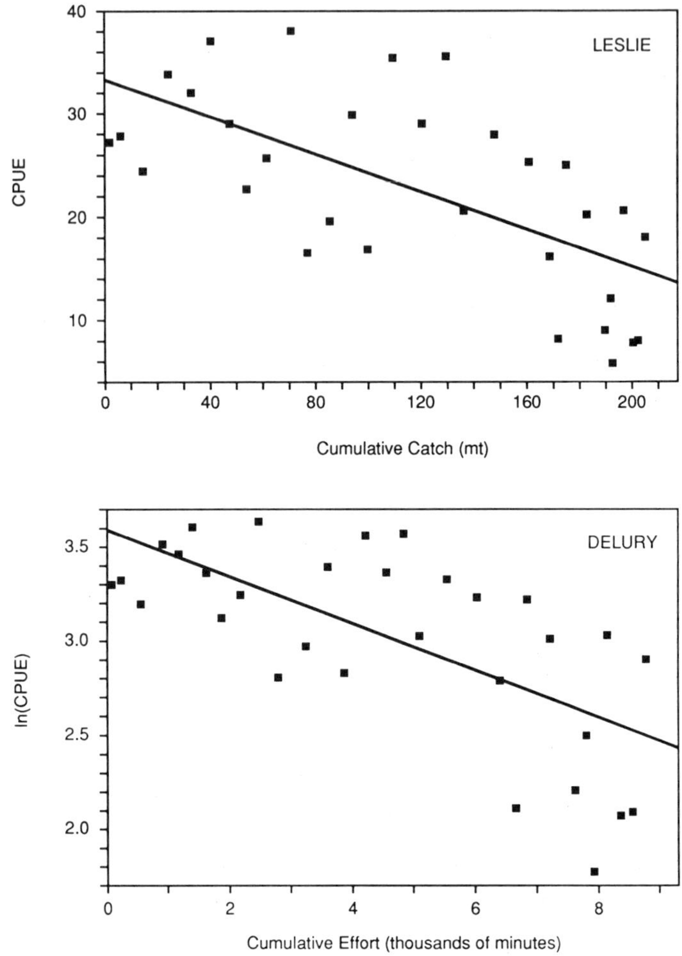

Figure 2-8. Relation between CPUE and cumulative catch and cumulative effort data used to obtain Leslie and Delury estimates of population size. From Iribarne et al. (1991).

depletion experiments. This minimized the impact of immigration and emigration on the results for scallops, although it was suspected that the results for prawns were still influenced by migration effects.

The resulting data were fit to a Leslie model, and since $a = A$ in this design, gear efficiency is equal to q. The results showed that the commercial prawn trawls (8 m wingspan) used in the experiment were 60–64% efficient for scallops. Trawl efficiency for western king prawns (*Penaeus latisulcatus*) differed significantly between the sexes, and between replicate experiments, the latter differences being attributed to differences in light levels.

Both tagging experiments and depletion experiments have been used to evaluate the efficiency of otter trawls used in surveys of king crab in the eastern Bering Sea. The width of the net while being towed (12–16 m) has been taken as the effective area swept for this species, and this appears to be reasonable in view of the king crab's limited locomotor capabilities, and the relatively large size of the trawls used. The 400 mesh eastern and modified eastern trawls used in these surveys were designed to ensure contact between the footrope and the bottom. Footropes are about 30% longer than the headropes and are weighted with chain (Otto 1986). Because of their size, few king crab probably escape beneath the footrope, so the efficiency was assumed to be 100% for those individuals in the path of the net.

Petersen mark–recapture experiments have been conducted to test the validity of this assumption, and have tended to give population estimates that agree well with those from area swept calculations. Otto et al. (1977) tagged 7431 legal-sized (greater than 164-mm carapace width) males in 1976, releasing them in such a manner as to affect a random distribution within the population. Of these, 35.4% were recaptured, so that the resulting Petersen estimate was quite precise, yielding approximate 95% confidence limits of 30.0 ± 1.2 million crab. This compared favorably with the 1976 survey estimate of 32.7 ± 5.2 million legal sized males. Hoopes and Greenough (1968) reported a similar situation in 1966, where the Petersen estimate yielded 95% confidence limits of 13.2 ± 1.1 million legal males, while the survey estimate was 11.3 ± 2.7 million.

Leslie population estimates were also made, using the commercial CPUE for the Bristol Bay fishery. This design proved less effective than those employing a smaller, more well-defined area for the depletion experiments, since the resulting Leslie estimates were consistently smaller than the survey estimates (Fig. 2-9). Differences between population estimates were largely explainable on the basis of lower exploitation rates on crab in the smaller legal size groups, and by the fact that areas with sparse concentrations of legal crab are included in the survey estimates, but are not fished commercially.

Gear Standardization and Intercalibration

It should be clear by now that a significant research effort should precede any trawl survey. Pilot surveys should be conducted to delineate geographic distribution, diel, and seasonal migration patterns of the target species. Gear and fish behavior studies should be conducted to determine whether net width or door spread is the most appropriate parameter in defining the effective area swept, and to estimate the efficiency of the gear in capturing the target species that lie in this critical zone. It should be decided which net parameters need to be monitored continuously during routine surveys. The gear should be tailored to minimize escapement under the footrope (through sweepline/bridle design, addition of tickler chains or rockhopper gear) and achieve optimal sweepline length, door spread, and vertical and horizontal net openings. Experiments should also be carried out to determine the optimal towing

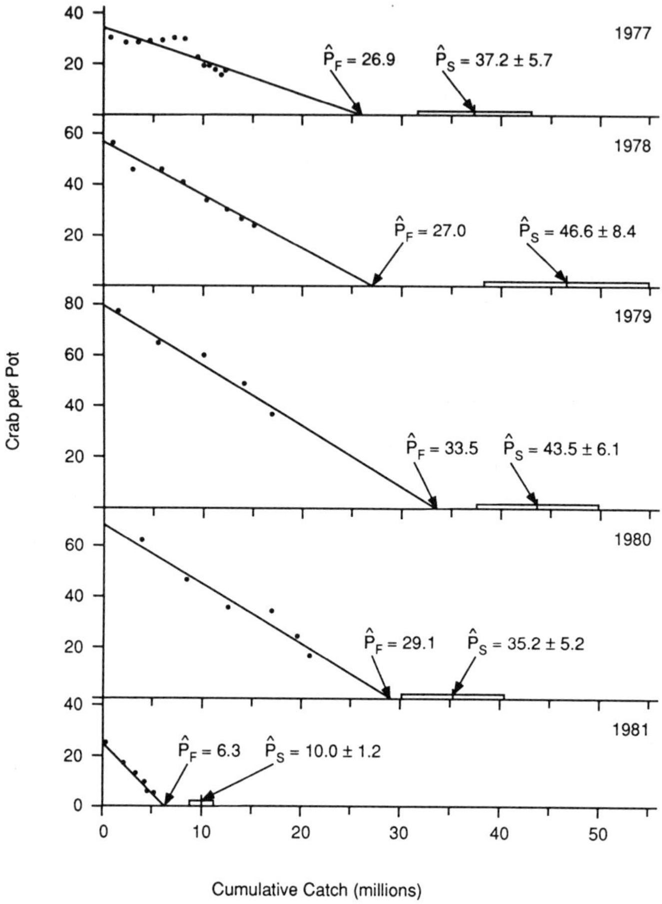

Figure 2-9. Comparison of fishery-derived Leslie estimates ($\hat{P}_F$) with trawl survey estimates ($\hat{P}_S$) of the population size for legal male red king crab in Bristol Bay, Alaska. Plotted points represent successive weeks in the fishery. After Otto (1986).

speed. Speeds that are either too slow or too fast to ensure stable gear performance or optimal herding by the target species should be identified.

Once this research is completed, all aspects of gear design and deployment should be carefully documented (Table 2-1). Quality control should be such that trawl design and deployment are carefully standardized and monitored during the course of all survey cruises.

Even with careful standardization and monitoring of gear performance, coefficients of variation around any density estimates will still remain high (Rose and Walters 1990; Koeller 1991), however, owing to the patchy spatial distributions that characterize most species. This high variability is a key consideration when the temptation to alter the survey methodology arises, since such changes always necessitate intercalibration or comparative trawling

Table 2-1. Documentation of trawl design and deployment

Trawl system	Vessel: length, tonnage, horsepower, navigation equipment
	Mensuration equipment: overall design, number and deployment of sensors
	Construction of the net: number of meshes per section, mesh size, twine thickness, twine material, and color; this information is normally given in a technical drawing
	Sweeplines and bridles: length, diameter, material, rigging; detailed description or technical drawing
	Headrope: length, diameter, material
	Footrope: length, diameter, material, extra weight (chain), how distributed
	Roller gear or tickler chain arrays (if applicable); detailed description or technical drawing
	Floats: number, volume, material, where attached
	Trawl doors: design, size, weight, material
Deployment	Towing speed
	Warp length and bottom depth
	Monitoring of performance: net width, vertical opening, door spread, time on/off bottom
	Season, time of day, light and current conditions
	Protocol for determining untrawlable bottom and locating alternate stations

experiments. While these experiments are conceptually simple, the high variances associated with them makes them costly to implement and result in fishing power correction factors that are characterized by wide confidence limits. Type II statistical error (probability of failing to detect differences between gear types given that differences actually exist) is quite high, and researchers should take little comfort in failing to detect differences between gear types unless a careful statistical power analysis (Peterman 1990) has been carried out.

One way to reduce the variances associated with fishing power comparisons is to block so as to incorporate "haul" or "station" effects into the experimental design. This can be accomplished by employing two or more vessels towing side by side or switching gear immediately after the tow has been made and completing a second tow in an area adjacent to the first. The reduction in error mean square (increase in statistical power) achieved by incorporating a "haul" effect into three-vessel intercalibration experiments in the eastern Bering Sea varied with species (as would be expected on the basis of their differing spatial distributions), ranging from 32 to 85% (Gunderson 1976a).

Significant differences often exist between different size groups (Gunderson 1976a, Hoff 1989), so that size effects should also be incorporated into the design. It is also important that the intercalibration hauls be made over a wide range of areas and conditions so that the results can be extrapolated to the entire survey area.

Assuming that a logarithmic transformation of the catch data will result in a normal distribution of errors (Robson 1966, Wilderbuer 1988), with mean zero and constant variance, a typical ANOVA design for intercalibration experiments would be

$$\ln(C_{ijk} + 1) = \mu + G_i + L_j + GL_{ij} + H_k + e_{ijk}$$

where C_{ijk} = catch per unit distance fished (number or weight) with gear i for fish of length j at haul location k
μ = grand mean
G_i = gear effect (fixed)
L_j = length effect (fixed)
GL_{ij} = gear × length interaction (fixed)
H_k = haul effect (random)
e_{ijk} = error

A maximum likelihood estimation program such as BMD 3V (Dixon et al. 1988) can be used to obtain direct estimates of model parameters and their variances. Length-specific gear correction factors can then be calculated as

$$Q_{ij} = \exp[(G_{ij} + GL_{ij}) - (G_{1j} + GL_{1j})]$$

where Q_{ij} = fishing power correction coefficient for length j and gear i relative to the standard (gear 1). The variance of this estimate can be obtained using the delta method (Seber 1973; Wilderbuer 1988).

Results from several intercalibration experiments using this general methodology have shown that the confidence intervals around fishing power coefficients are typically quite wide, even when a large number of comparative tows have been made. Wilderbuer (1988) made 71 pairs (142 total) of comparative hauls in order to examine the effect of changing the standard Nor'-eastern trawl used in the Gulf of Alaska. Due to extensive gear damage on rough bottom, it was necessary to change the construction material from nylon to polypropylene and to remove 19.5 feet of webbing from the lower panels to create a "flying wing." Even with 71 pairs of tows, however, the confidence limits around the fishing power coefficient for most species were quite broad, and usually included 1.00 (implying no significant differences between gear types). These confidence limits were asymmetric due to the logarithmic scale and were obtained by multiplying (upper confidence limit) or dividing (lower confidence limit) the point estimate by a factor that ranged from 1.08 for arrowtooth flounder to 1.89 for rougheye rockfish (Table 2-2).

Hoff (1989) reported on a series of 26 pairs of tows designed to evaluate the effect of attaching an extra 61 cm of chain between each end of the footrope and dandyline of a "Modified Eastern" otter trawl used in the eastern

Table 2-2. Fishing power coefficients (95% confidence limits in parentheses) estimated by intercalibration experiments in the Gulf of Alaska (Wilderbuer 1988) and eastern Bering Sea (Hoff 1989)

Species	Wilderbuer (1988)
Arrowtooth flounder	0.92 (0.85–0.99)
Halibut	0.88 (0.75–1.02)
Flathead sole	0.94 (0.73–1.20)
Rex sole	0.95 (0.70–1.23)
Dover sole	0.91 (0.83–1.00)
Pollock	0.96 (0.85–1.09)
Sablefish	0.94 (0.81–1.10)
Pacific cod	0.97 (0.83–1.13)
Pacific ocean perch	0.86 (0.71–1.05)
Thornyheads	0.84 (0.69–1.02)
Shortraker rockfish	1.09 (0.77–1.55)
Rougheye rockfish	1.15 (0.61–2.17)

	Hoff (1989)	
Length (cm)	Yellowfin sole	Rock sole
16–20	9.88 (4.04–24.12)	4.26 (2.75–6.61)
21–25	3.29 (1.66–6.54)	3.50 (2.26–5.42)
26–20	2.39 (1.29–4.42)	2.73 (1.76–4.23)
31–35	1.94 (1.04–3.60)	2.08 (1.34–3.24)

Bering Sea. These gear changes were made to improve bottom contact. At the same time, 55-m double bridles running from the wings to the doors were used to replace the former rigging consisting of a single 46-m cable from the door, which then branched into two 27-m bridles attached to the wings. Hoff found significant differences between the original and improved designs, with fishing power coefficients ranging from 1.94 for 31- to 35-cm yellowfin sole to 9.88 for 16- to 20-cm yellowfin sole (Table 2-2). Once again, however, the

95% confidence intervals encompassed a wide range, the factor used to determine the confidence limits ranging from 1.55 to 2.44.

Intercalibration experiments are a clear example of a technique characterized by low statistical power. Under most field conditions these experiments are capable of detecting statistically significant differences between gear types only if there are gross differences in catch rates. Furthermore, the investigator is frequently forced to standardize catches using point estimates that can be relatively large, yet imprecise. Using Hoff's results, for example, it would be necessary to multiply catches of 16- to 20-cm yellowfin sole made with the "standard" net by a factor of 9.88 to compare them with those obtained using the improved footrope design. This factor could range from 4.04 to 24.12 using the 95% confidence limits, however, putting the analyst in the unhappy position of making substantial changes in the data base, yet having little confidence in the accuracy of the results.

This situation is not uncommon, even when a large number of comparative tows are made, and should be avoided in any manner possible. Extensive gear research, development, and testing should precede any routine series of surveys, rather than being conducted (however inadvertently) during the surveys themselves.

SAMPLING THE CATCH

The main parameters of interest during a trawl survey are population size and/or biomass estimates for selected target species. A second goal of most trawl surveys is to determine the length and/or age composition of these same species, by sex. Trawl surveys offer an opportunity to undertake a number of ancillary collections, such as data on individual weights, maturity, fecundity, stomach contents, and tissue samples for genetic, pathology, or contaminant studies. Survey vessels also offer a convenient platform from which hydrographic or water chemistry studies can be carried out.

The primary objective is to estimate population sizes, however, and it is imperative that a large number of stations be sampled to do this properly. As a result, ancillary studies should be avoided unless they can be carried out without slowing down the progress of the survey vessel. If the vessel must frequently wait for the scientific complement to complete the processing of the catch before it can bring in the next trawl haul, the researchers must either add more people to the scientific staff, or scale back the amount of data collected. This is clearly a problem in evaluating the tradeoffs involved in available funds, personnel, time, and space, and the solution will vary from survey to survey.

Time is always of the essence in processing trawl catches, and a number of subsampling procedures have been developed over time. These procedures were derived empirically by taking subsamples in some prescribed manner, sampling several catches in their entirety, then using χ^2 tests to evaluate the agreement of parameters obtained from the subsamples with those for the entire catch.

Species Composition

Because of the tendency of fish to stratify in the trawl cod-end on the basis of buoyancy changes during decompression (e.g., swim bladder distention), subsampling catches for species composition presents serious problems. Debris and aquatic vegetation in the net often complicate the subsampling problem further. Fortunately, catches are often small enough to sort in their entirety, provided that too much time is not spent in detailed itemization of every species. While it is imperative that catches of the target species be carefully enumerated for every haul, no matter how few are caught, liberal use of a miscellaneous category should be made for species of lesser importance.

The catch should be sorted into baskets by species (or a "miscellaneous" category) and weighed. It is not necessary to weigh each basket, as it has been shown that the catch per sampling basket or tub for a given species varies relatively little (Westrheim 1967, 1976). As such, it is necessary to weigh only about six sampling baskets to obtain a mean basket weight, and simply count the remaining baskets as they are discarded over the side of the vessel. Only those baskets that will be required for length and age subsamples need to be retained. Westrheim (1976) showed that the error introduced by using this technique was always less than 8%, and averaged 0.6% during 108 subsampling studies involving a variety of species (including cod, rockfish, and flatfish). This approach has the added advantage of greatly reducing congestion on the deck, as well as the handling mortality for those fish that are eventually discarded.

Extremely large catches (greater than 1 mt) can take several hours to process if they are sampled in their entirety, and Hughes (1976) developed a technique for subsampling them. His method takes into account the fact that fish will stratify vertically in the cod-end of the net during fishing and retrieval, those with swim bladders floating to the top while others sink. The procedure is implemented as follows:

1. A cargo net is placed in the deck bin where the catch is to be dumped. This will be used to remove a subsample, which is then processed using the procedures employed routinely for entire catches.
2. Prior to dumping the catch into the deck bin, the cod-end is turned so that the segments corresponding to the "top" and "bottom" during trawling (identified by marks that have been placed on the net) are positioned athwartships (Fig. 2-10).
3. If a suitable load cell or scale is available, the entire catch can be weighed prior to dumping. The weight of fish in the cargo net will be obtained during routine processing of its contents, and can be used to determine the proportion

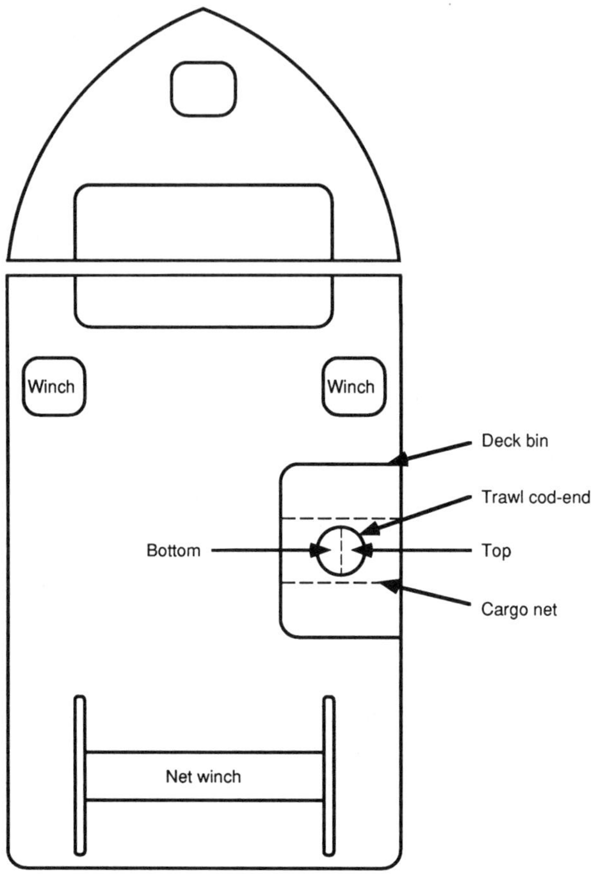

Figure 2-10. Positioning the cod-end to obtain a cargo net subsample of a large trawl catch.

of the catch sampled. Weighing the entire catch can prove difficult in rough seas, however, and as an alternative the proportion of the catch contained in the cargo net can be calculated from the volume of fish in the deck bin before and after the subsample has been removed.

Tests of the accuracy of this technique have been conducted by sorting and weighing entire catches (Hughes 1976, Gunderson 1976b), and have shown that errors in the estimate of percent composition are less than $\pm 10\%$ for the dominant species in the catch. Errors greater than $\pm 25\%$ were usually restricted to minor species comprising less than 5% of the catch. Length measurements were made for a variety of flatfish, roundfish, and rockfish species during these experiments, and showed that the size composition in the cargo net was representative of the total catch.

Size Composition

There is usually a tendency for samplers to take the largest fish first in the process of sorting the catch, and this should be taken into account when drawing a subsample of the catch for length or age samples. Strictly random selection of fish in the subsample is usually a poor idea, and a number of subsampling studies (Westrheim 1967; Gunderson 1976b) have shown that a representative subsample can be obtained most efficiently by taking an equal portion of the subsample from the first, middle, and last part of the catch sorted. An attempt should be made to use containers small enough so that entire baskets are the sampling unit, but when the catch is comprised of small fish it may prove necessary to take only the top fraction of a single basket from the first, middle, and last part of the catch to obtain a sample of the right size. Studies employing these techniques showed statistically significant (0.05 level) size composition differences between the subsample and the catch it was drawn from in only 1 of 22 catches studied by Westrheim (1967) and 1 of 11 studied by Gunderson (1976b), involving a variety of roundfish and flatfish.

Since the sample sizes used to estimate the length composition of each haul are roughly the same, a sample taken from a large catch will receive the same weight as one from a small one unless they are weighted appropriately. When estimating the length composition for a stratum, the fraction (f_i) of fish in sex-length group i should consequently be estimated from

$$f_i = \frac{\sum_{j=1}^{n} C_j f_{ij}}{\sum_{j=1}^{n} C_j}$$

where f_{ij} = fraction of sex-length group i in haul j
C_j = catch per unit distance fished for haul j
n = number of hauls

Age Composition

It is seldom necessary to draw a subsample larger than 200 fish for sex and length observations, or 100 fish from which structures (scales, otoliths, fin rays, etc.) are removed for age determination. Several studies (Kimura 1977, Lai 1987) have shown that random age subsamples are more efficient statistically than those obtained by taking age structures from a fixed number in each length group (e.g., 5 from each sex–cm group), so that it is most convenient to subsample for length and age simultaneously. For example, if the goal is to obtain sexes and lengths from 200 fish, and age structures from 100 of these, two subsamples of 100 each could be drawn, taking an equal fraction of each from the first, middle, and last part of the catch. The weights of all

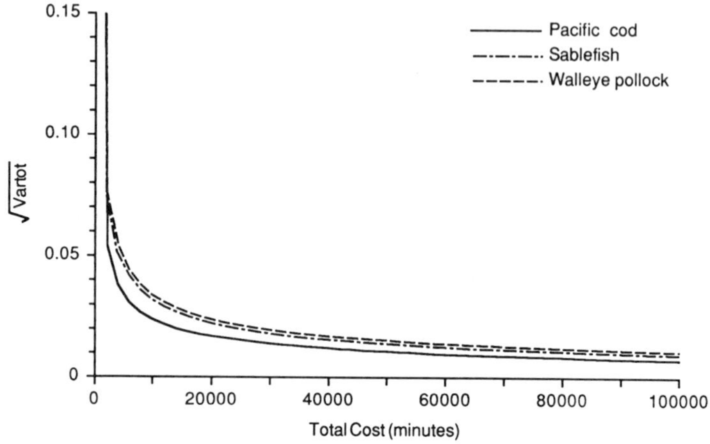

Figure 2-11. The relationship between $\sqrt{\text{Vartot}}$ and total cost (minutes) for Pacific cod, sablefish, and walleye pollock, using random age subsampling in an age–length key. After Lai (1987).

subsamples should be carefully recorded, as they will be required later to extrapolate the results to the entire catch.

Because of the amount of effort required to age fish properly, and document the validity of these ages, it will seldom be cost effective to collect age determination structures from every haul. Depending on the species, no more than about 500–1500 age observations should be required during the entire survey when using an age–length key (Lai 1987). Using Vartot (the sum of the variances in the estimated proportion at each age) as a criterion, Lai showed that precision increased sharply with increases in allotted time until 10,000 minutes was reached (Fig. 2-11). This corresponded to about 70 working days, and represented a point of diminishing returns for the three species he studied. Lai showed further that optimal allocation of the 10,000 minutes in an age-length key would require approximately 2000 length observations and 800 random age observations for Gulf of Alaska sablefish, 2500 lengths and 1200 ages for eastern Bering Sea walleye pollock, and 3000 lengths and 600 ages for Pacific cod (*Gadus macrocephalus*) from the Washington coast.

As a result a reasonable goal in any survey will be to obtain a length–sex sample for the target species in every haul, using the results to estimate the length composition of the total population, by sex. Subsamples from which aging structures are taken can be drawn from a relatively small percentage of these hauls and used to estimate age composition using an age–length key. It is important that the hauls from which age structures are drawn are representative of the entire survey area, however.

DATA RECORDING AND EDITING

Examples of data forms that can be used during a trawl survey are shown in Figures 2-12 and 2-13. A waterproof form should be developed for use on

SPECIES CATCH FORM

VESSEL [1 2 3] CRUISE [5 6 7] HAUL [9 10 11]

YOUR NAME ____________

SPECIES NAME	CODE (13 14 15 16 17)	WEIGHT TO A TENTH LB (19 20 21 22 23 . 25)	NUMBER (27 28 29 30 31 32)	COMMENTS
		.		
		.		
		.		
		.		
		.		
		.		
		.		
		.		
		.		
		.		
		.		
		.		
		.		
		.		
		.		

LENGTH-FREQUENCY FORM

VESSEL [1 2 3] ____________ CRUISE [5 6 7] HAUL [9 10 11]

DATE ____________ TOTAL WEIGHT (LB.) ____________ SUBSAMPLE WEIGHT (LB.) ____________

SPECIES CODE [13 14 15 16 17] SPECIES NAME ____________

YOUR NAME ____________ SUBSAMPLE TYPE [32] LENGTH TYPE (IF NOT FORK LEN.) [33 34]

LENGTH IN MM. (19 20 21 22 23)	FREQ-UENCY (26 27 28)	SEX (30)	LENGTH IN MM. (19 20 21 22 23)	FREQ-UENCY (26 27 28)	SEX (30)	LENGTH IN MM. (19 20 21 22 23)	FREQ-UENCY (26 27 28)	SEX (30)
0			0			0		
0			0			0		
0			0			0		
0			0			0		
0			0			0		
0			0			0		
0			0			0		
0			0			0		
0			0			0		

Figure 2-12. Examples of forms used to record data on catch by species and length–sex composition. Numbers above blocks indicate column to be used during data entry.

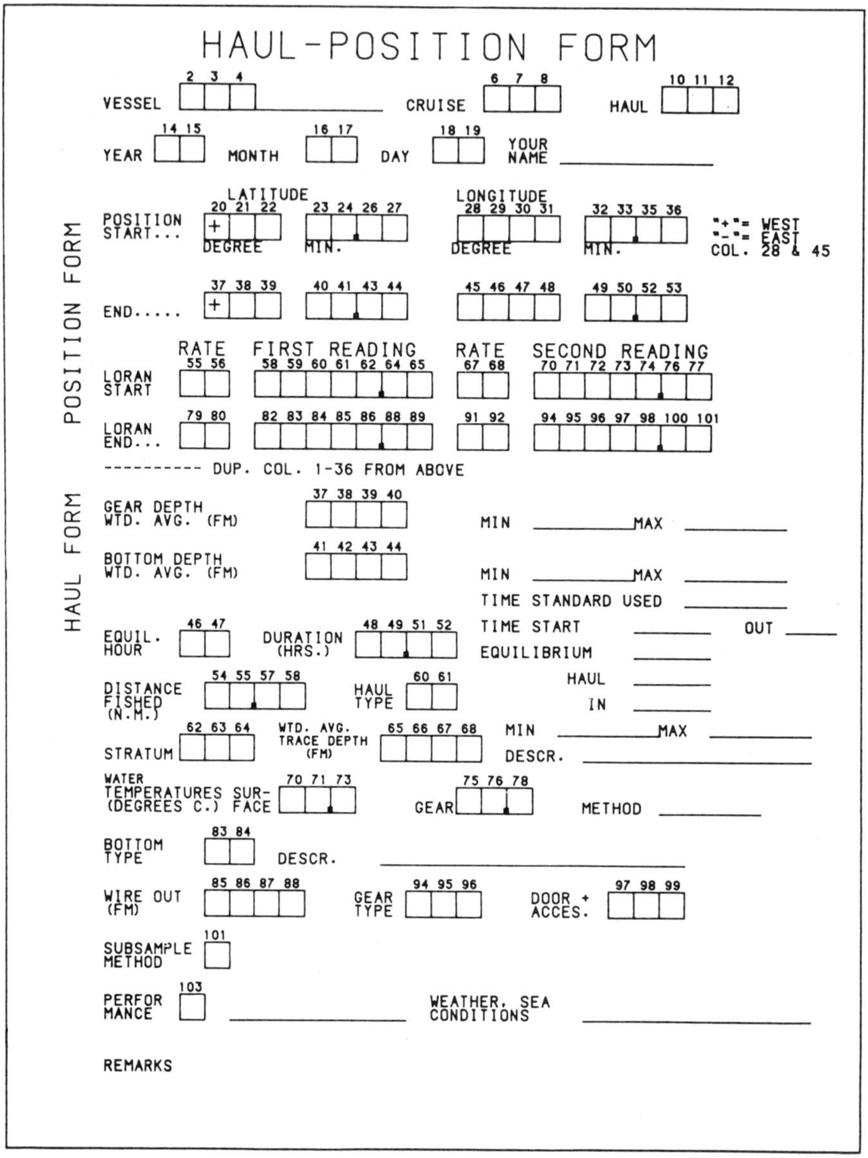

HAUL-POSITION FORM

VESSEL [2 3 4] ____ CRUISE [6 7 8] HAUL [10 11 12]

YEAR [14 15] MONTH [16 17] DAY [18 19] YOUR NAME ____

POSITION FORM

LATITUDE / LONGITUDE

POSITION START... [20 21 22] DEGREE [23 24 26 27] MIN. [28 29 30 31] DEGREE [32 33 35 36] MIN. "+"= WEST "-"= EAST COL. 28 & 45

END..... [37 38 39] [40 41 43 44] [45 46 47 48] [49 50 52 53]

RATE / FIRST READING / RATE / SECOND READING

LORAN START [55 56] [58 59 60 61 62 64 65] [67 68] [70 71 72 73 74 76 77]

LORAN END... [79 80] [82 83 84 85 86 88 89] [91 92] [94 95 96 97 98 100 101]

---------- DUP. COL. 1-36 FROM ABOVE

HAUL FORM

GEAR DEPTH WTD. AVG. (FM) [37 38 39 40] MIN ____ MAX ____

BOTTOM DEPTH WTD. AVG. (FM) [41 42 43 44] MIN ____ MAX ____

TIME STANDARD USED ____

EQUIL. HOUR [46 47] DURATION (HRS.) [48 49 51 52] TIME START ____ OUT ____

EQUILIBRIUM ____

DISTANCE FISHED (N.M.) [54 55 57 58] HAUL TYPE [60 61] HAUL ____ IN ____

STRATUM [62 63 64] WTD. AVG. TRACE DEPTH (FM) [65 66 67 68] MIN ____ MAX ____ DESCR. ____

WATER TEMPERATURES (DEGREES C.) SUR-FACE [70 71 73] GEAR [75 76 78] METHOD ____

BOTTOM TYPE [83 84] DESCR. ____

WIRE OUT (FM) [85 86 87 88] GEAR TYPE [94 95 96] DOOR + ACCES. [97 98 99]

SUBSAMPLE METHOD [101]

PERFOR MANCE [103] ____ WEATHER, SEA CONDITIONS ____

REMARKS

Figure 2-13. Example of a form used to record data on the vessel, gear, time, depth, location, etc., for each haul. Numbers above blocks indicate column to be used during data entry.

deck. This need only contain the raw data for each haul, consisting of basket weights and the number of baskets counted for each species or species group, together with subsample weights for those specimens selected for length–sex measurements. The raw data from the on-deck form should be retained in the event they are required later to resolve the source of an error.

At convenient intervals, the data from the on-deck forms should be used to obtain estimates of the total catch of each species or species group, and the results transcribed onto permanent forms (Fig. 2-12). If a microcomputer is available, a preferable approach would be to enter the data directly into a disk or permanent file. Programs can be written to carry out the calculations required to estimate catch sizes from the raw data (minimizing some of the errors that creep in when carrying out even simple calculations at sea or after a hard day in the field), and some preliminary error checks incorporated. At the very least, acceptable ranges can be defined for lengths and species codes, so that some obvious errors can be caught immediately.

Length–sex observations are best tallied on acetate sheets that have been scribed at appropriate (1 mm, 1 cm) intervals. Preferably, these sheets can be detached from the measuring board, stored, and the resulting length frequency data transcribed onto permanent forms (Fig. 2-12) or entered into a microcomputer at a convenient time. Errors in proper recording and extrapolation of subsample sizes are among the most common encountered, and can result in significant flaws in the data if undetected.

An alternative method of recording and logging length data is to use an electronic measuring board that uses a probe to record the fish's length on a sensitive Cartesian grid, or read a bar code corresponding to a given length. These devices are capable of temporarily storing the length frequency information and downloading it to a host computer. Such systems are commercially available and at least one model can be interfaced with digital calipers for measuring and storing data on carapace dimensions of shellfish.

On a large research vessel, most of the information on the haul-position form (Fig. 2-13) is most readily recorded by the vessel captain or someone else familiar with the operation of the navigation equipment. Water temperature and salinity data for each haul can be recorded on the on-deck form, and added to the haul position form (or entered into a computer file) later.

A data checking program should be developed to specify acceptable ranges for the raw data (position, depth, distance fished, haul duration, etc.) on the haul-position form. A number of other parameters can be calculated and screened for errors. The most important of these is the estimated towing speed, obtained by dividing distance fished by haul duration. This parameter should fall within a relatively narrow range, and any outliers can warn the person inputting or checking the data of potential errors in either positioning or times. The sooner these checks are carried out, the easier it is to correct erroneous entries. Unless data entry is carried out during the survey itself, however, it is often impossible to resolve errors in navigation and positioning, and the investigator will be forced to discard data from questionable hauls.

EXAMPLES OF SUCCESSFUL APPLICATIONS

Northwest Atlantic

The multispecies groundfish surveys carried out in the northwest Atlantic are probably the ones most familiar to fisheries scientists and managers. Similar methodology has been adopted by United States scientists for surveys of the continental shelf from Jacksonville, Florida, to Nova Scotia, and by their Canadian counterparts from Nova Scotia to Newfoundland (Doubleday and Rivard 1981). These surveys are all based on a stratified random statistical design (Fig. 2-14), with strata being established on the basis of bathymetric and geographic limits.

The bathymetric strata follow logically from the changes in the groundfish community with depth, and the well-demarked bathymetric limits for the distribution of the target species. On the Grand Bank, for example, the 50-fm contour effectively delineates the maximum depth for yellowtail flounder, and 150 fm represents an upper limit to the distribution of American plaice. Cod have a relatively wide depth range, extending to at least 200 fm in some parts of the Bank (Pitt et al. 1981). The basis for geographic subdivision of these strata is more arbitrary, based primarily on the need to spread sampling effort out in a relatively uniform manner. Oceanographic and ecological domains are also taken into account in the delineation of these strata, however.

While all of these surveys are described as "stratified random," the selection of stations is not necessarily a strictly random process. In the case of the United States surveys, for example, no more than one station in a given 5 minute latitude by 10 minute longitude square is selected, ensuring a dispersion of stations (Azarovitz 1981). Allocation of stations between strata is roughly proportional to area (rather than the variance of the catches of target species) in the United States surveys, further ensuring adequate dispersion of sampling effort.

Many of the northwest Atlantic surveys have adopted a protocol of surveying throughout all 24 hours of the day. However, it has become clear that several of the target species show significant diel changes in abundance (Pitt et al. 1981; Walsh 1991), so that the results are subject to interannual variability in both fish behavior and the diel distribution of the trawl sampling. Surveys in some areas are restricted to daylight hours (Forest and Minet 1981), apparently in recognition of this problem.

In most of the Northwest Atlantic surveys, a standard tow is typically made for a constant time (e.g., 30 minutes) at a target towing speed (e.g., 3.5 knots). The actual distance fished during a standard tow can be highly variable, however. In cruises where a Doppler speed log was used, speed over ground was observed to range from 0.8 to 5.0 knots during United States surveys where the target speed was 3.5 knots (Byrne et al. 1981). A standard "#36 Yankee" trawl (10.4–10.9 m horizontal mouth opening) has been used for the United States' autumn surveys since 1963, although designs with and without roller gear on the footrope are employed in different areas (Azarovitz

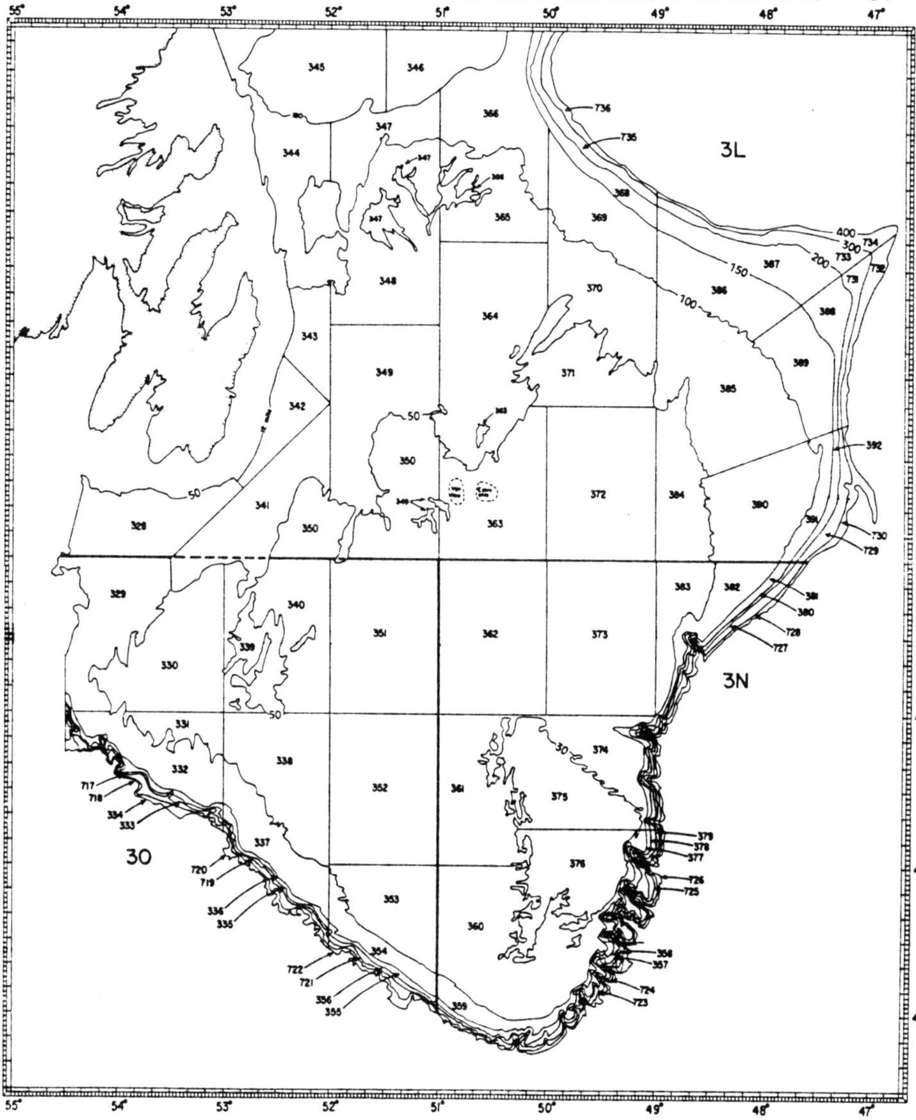

Figure 2-14. Stratification used for a trawl survey of the Grand Bank. Depth contours in fathoms. After Pitt et al. (1981).

1981). Approximately 300 stations are sampled during each United States survey, and since 1979 trawl performance has been monitored using headrope transducers or other mensuration equipment.

The definition of the standard tow on the basis of trawl duration rather than distance fished, together with the 24-hour survey protocol both tend to increase the variance around the estimates of mean catch per tow. As such, estimates of precision for these surveys could probably be reduced substantially. The 95% confidence intervals around survey means obtained for yellowtail flounder on the Grand Bank ranged from ± 30 to 56% during 1971–

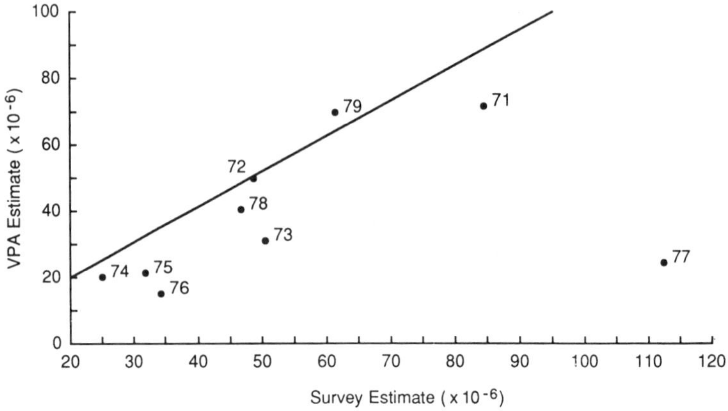

Figure 2-15. Plot of annual yellowtail flounder population estimates (number of fish 7 years and older) from virtual population analysis (VPA) versus trawl surveys for selected strata in NAFO (Northwest Atlantic Fisheries Organization) statistical areas 3L and 3N. After Pitt et al. (1981).

1980 (Pitt et al. 1981), while similar values for the most important species in United States surveys (cod, haddock, and yellowtail flounder) typically range around ±50% (Byrne et al. 1981).

Area-swept estimates of yellowtail flounder abundance have been calculated for the Grand Banks, and show good agreement with results from catch-age analysis (Fig. 2-15). As such, the survey estimates provide information on the current size of the stock that is of value to managers (Pitt et al. 1981).

The stratified mean catch per survey tow has been used as a relative index (Clark 1981; Pennington 1986) of abundance for stocks along the northeast coast of the United States, eliminating the need to make strict assumptions about area swept per tow or availability of target species. In the case of haddock, trends in the relative indices compare well with independent results from catch-age analysis (Fig. 2-16). These relative indices have often been used in conjunction with catch-age analyses to develop time series of absolute abundance (Clark 1981; Collie and Sissenwine 1983), for species such as yellowtail flounder and haddock that have proven to be of considerable importance in the management of these stocks.

Eastern Bering Sea

The continental shelf in the eastern Bering Sea is relatively smooth, and ideally suited to trawl surveys. Untrawlable bottom is uncommon, eliminating the dilemma faced in areas where rough, rocky areas abound. As pointed out by Pope (1988), "Random tows have an uncanny habit of coming on top of rough ground which cannot be worked and the subsequent discussion with a fishing skipper on where to work is always bias-prone." Even when the adjacent area is searched until trawlable bottom is found, or a secondary station is

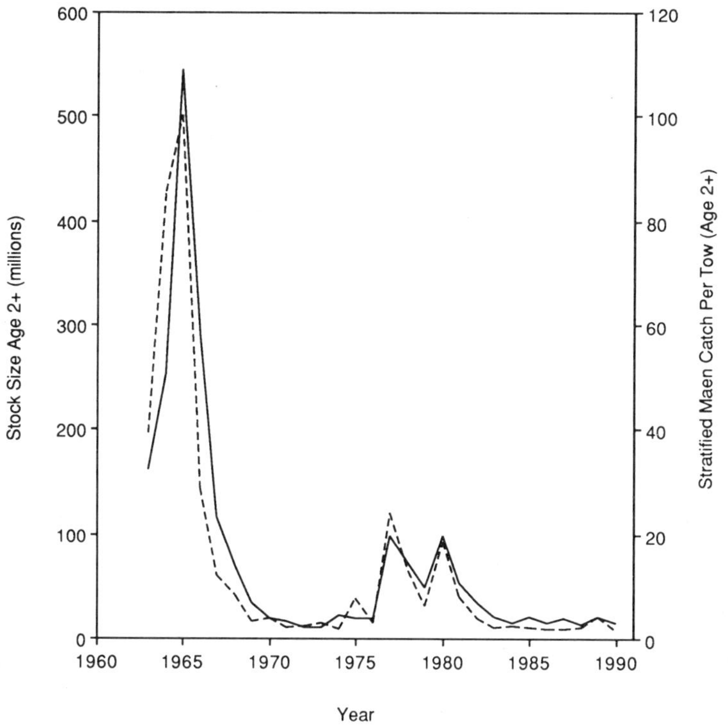

Figure 2-16. Abundance of haddock 2 years and older on Georges Bank, based on virtual population analysis (VPA, solid line) and trawl surveys (dashed line). After Hayes and Buxton (1992).

randomly chosen in advance as a backup, serious questions still remain as to how the resulting data should be treated.

On one hand, the investigator can assume that densities over untrawlable bottom are similar to those over trawlable bottom, and simply extrapolate the survey catches from trawlable bottom to the entire survey area. An equally valid assumption, however, is that the groundfish community changes completely over untrawlable bottom, and that such areas can be deleted from consideration when estimating abundance. In either case serious bias could result unless the validity of the underlying assumptions is checked through use of camera surveys or submersibles.

The Bering Sea surveys are designed using a systematic grid, with trawl stations located at the center of each grid square (Fig. 2-17). Grid size varies between strata, with highest station densities being allocated to areas where target species are most abundant. In some years, additional stations have been placed at the corners of the grid squares to increase precision for areas of special importance.

Current practice is to conduct the survey using two vessels, each vessel sampling alternate north–south transects. The survey progresses in a westerly

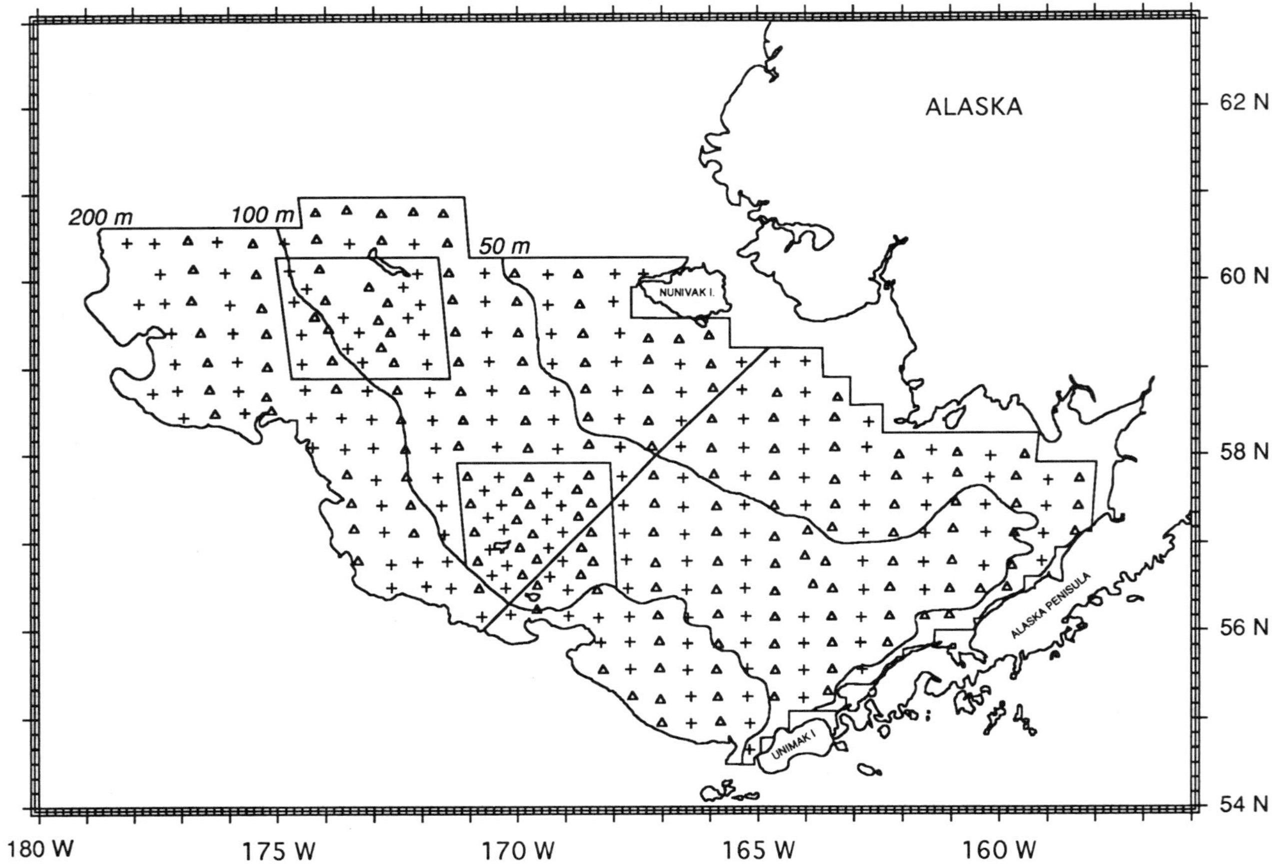

Figure 2-17. Sampling grid used during trawl surveys of the eastern Bering Sea in 1991. Solid lines denote stratum boundaries based on depth, latitude (North or South), and station density. Triangles and crosses denote stations sampled by each of the two vessels participating in the survey. Courtesy of G. Walters, National Marine Fisheries Service.

direction, beginning in Bristol Bay in June, so that the vessels move counter to the onshore feeding migration by yellowfin sole and other flatfish. Delays in the progress of the survey and changes in survey timing are avoided, since they can introduce considerable bias into survey results (Hoff 1989).

The surveys began in 1969, and were initially designed to monitor the abundance of king crab. The area covered has expanded geographically over the years as additional target species were added. The surveys currently provide population or biomass estimates for king crab (*Paralithodes* spp.), tanner crab (*Chionoecetes* spp.), yellowfin sole (*Limanda aspera*), rock sole (*Lepidopsetta bilineata*) , Alaska plaice (*Pleuronectes quadrituberculatus*), and other flatfish that play an integral role in the management of these stocks. The surveys also provide indices of the relative abundance of species such as cod and pollock that are used as auxiliary information in catch-age analyses.

A standard "400-Mesh Eastern" otter trawl (12–14 m wing spread) was used during 1969 through 1980, and was replaced with a larger version of this design ("83-112," 14–16 m wing spread) during 1982 to accommodate the larger vessels being used in the surveys. A series of gear and comparative trawling cruises conducted during 1981 served to optimize the design of the new trawl. A number of trawl mensuration studies have been carried out over the years, and trawl wing spread has been monitored on most tows since 1988.

A standard tow is conducted by towing at a fixed target speed (over ground) of 3 knots for 30 minutes. The distance fished during each tow is then measured directly using navigational fixes at the beginning and end of each tow, and these are used in conjunction with estimates of trawl wing spread to estimate the area swept for each tow. Biomass or population estimates and their variance are then obtained using Equations 1.1 and 1.2.

Comparative trawling experiments (Otto 1986) and efficiency experiments using tagging or depletion experiments (see Estimating Gear Efficiency) suggest that changes in gear have had little effect on the catch per area swept for king crab, presumably due to the limited range of avoidance responses that this species is capable of. One researcher has characterized surveying king crab as being analogous to "scooping up hockey pucks," so that the operational net width of the survey trawls provides a realistic measure of the effective area swept.

The pecision of the population estimates for legal-size red king crab (*Paralithodes camtschatica*) is relatively high, with 95% confidence limits for the major stock ranging from ±10 to 27% of the estimate during 1969–1983 (Otto 1986). These figures may overstate the precision to some extent since the data were sometimes poststratified, with strata being delineated on the basis of trends in abundance observed during the survey. King crab population estimates have provided the sole basis for the management of this important stock, since age determination is not possible.

The situation is more complex for many species of groundfish since significant differences in efficiency, varying substantially between different spe-

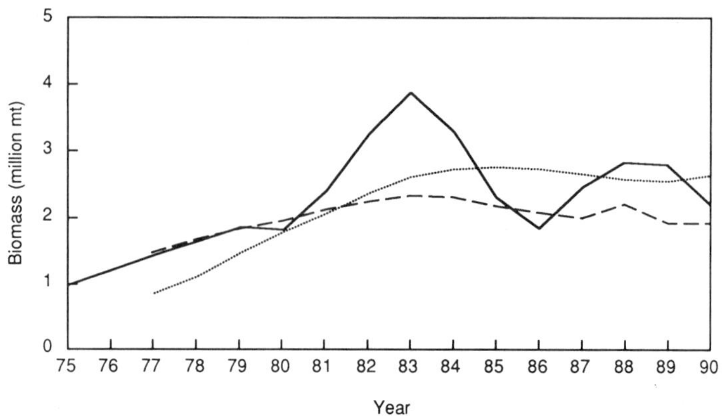

Figure 2-18. Estimated exploitable biomass (mt) of yellowfin sole in the eastern Bering Sea derived from trawl surveys (solid line), cohort analysis (dashed line), and a stock synthesis model (dotted line). After Bakkala and Wilderbuer (1991).

cies and size groups, accompany changes that are made in the trawls or their rigging (Gunderson 1976a; Hoff 1989). Biomass estimates have been made for most important groundfish species, using trawl wing spread to compute area swept. While no directed efficiency studies comparable to those carried out for king crab have been attempted, the survey biomass estimates obtained for yellowfin sole (Fig. 2-18) and Alaska plaice show good agreement with independent estimates from catch-age analysis (Zhang and Sullivan 1988; Bakkala and Wilderbuer 1991). The precision of the biomass estimates for yellowfin sole is relatively high, with 95% confidence intervals ranging from ±12 to 17% during 1975–1991 (Bakkala and Wilderbuer 1991). Poststratification was never practiced with the yellowfin sole data, and all strata were established prior to the survey.

Estuarine-Nearshore Surveys

A number of trawl surveys have been carried out using small research vessels in estuarine and shallow nearshore waters. These surveys employ vessels ranging from rubber dinghies to 20-m vessels, depending on the depths fished and environmental conditions that are encountered. Small beam trawls (2–4 m, Fig. 2-5) are frequently used for these surveys, and the most appropriate target species are juvenile flatfish and crustaceans such as winter flounder (*Pseudopleuronectes americanus*, Pearcy 1962), plaice (Riley and Corlett 1965, Edwards and Steele 1968, van der Veer 1986), Dungeness crab, or English sole (Gunderson et al. 1990).

The design for these surveys usually consists of a systematic series of tracklines with trawl stations at fixed depths or locations, or a single tow (perpendicular to the shore) encompassing the entire trackline. Gunderson et al. (1990) employed a series of systematic tracklines in their nearshore surveys of English sole (*Parophrys vetulus*) and Dungeness crab (*Cancer magister*), but used a stratified random design within estuaries (Fig. 2-19). The station pattern selected is usually such that it can be completed in a few days, making it possible to complete several surveys a month if desired.

Distance fished is obtained using odometer wheels mounted on the trawl, through navigational fixes taken at the beginning and end of each tow, or through use of an optical range finder to determine the distance between floats marking the beginning and end points. Trawl width can be readily determined for a beam trawl, making calculation of the area swept relatively straightforward.

Gear design and development studies are simplified by the ability to carry out SCUBA diving at operational depths. Gear efficiency studies are also relatively simple to carry out in nearshore surveys, and have included the use of tagging experiments and enclosure studies (Edwards and Steele 1968), quantitative studies of vessel and trawl avoidance (Kuipers 1975), and day–night comparisons (Gunderson and Ellis 1986).

The data from these surveys have been used extensively in studying recruitment, mortality, and growth processes for juvenile fish and shellfish, and have provided considerable insight into their early life history. Precision for surveys of juvenile plaice in the Wadden Sea was relatively high during June

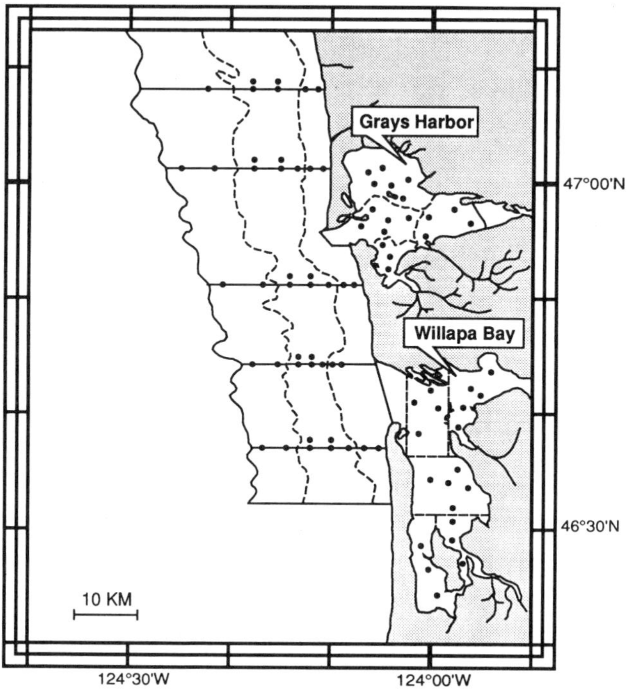

Figure 2-19. Sampling design employed in surveys of Washington estuaries and the adjacent coast. Dots represent sampling stations, solid lines represent coastal transects.

to September when the population had stabilized, with 95% confidence around density estimates being on the order of ±35% (Zjilstra et al. 1982). Approximate 95% confidence intervals around September population estimates for English sole in Washington estuaries and the adjacent open coast ranged from ±31 to 60% during the 5 years studied (Gunderson et al. 1990).

3

ACOUSTIC SURVEYS

INTRODUCTION

Because sound waves are capable of penetrating much greater distances than radio or light waves in water, sonar has become the preferred method of underwater observation where distances of more than about 30 m are involved. Sonar development has proceeded at a rapid pace since its initial use in antisubmarine warfare during World War I, and both the range and sophistication of sonar applications continue to increase. While much of this development has been carried out for defense purposes, it has revolutionized the manner in which commercial fisheries are carried out, and the ability of fisheries scientists to make quantitative observations of fish behavior and abundance.

Acoustic sampling differs from most other direct sampling techniques in that it offers the potential for almost continuous sampling along a series of transect lines. Several acoustic transmissions are typically made per minute, so that the volume being sampled can be conceptualized as a series of cones created by the spreading acoustic energy as the survey vessel proceeds along a preselected transect line (Fig. 3-1).

The physics of sound generation, reflection of echoes back toward the survey vessel, and conversion of these returning echoes to electrical energy by the processing equipment is well understood, and can be used to define the area being sampled with a high degree of accuracy. The same theory can be used to quantify the level of acoustic energy in the returning echoes, and if a single species is present, this can readily be used to estimate their total number or weight per unit volume. Acoustic systems used in standard fisheries surveys are not yet capable of discriminating the energy returning from dif-

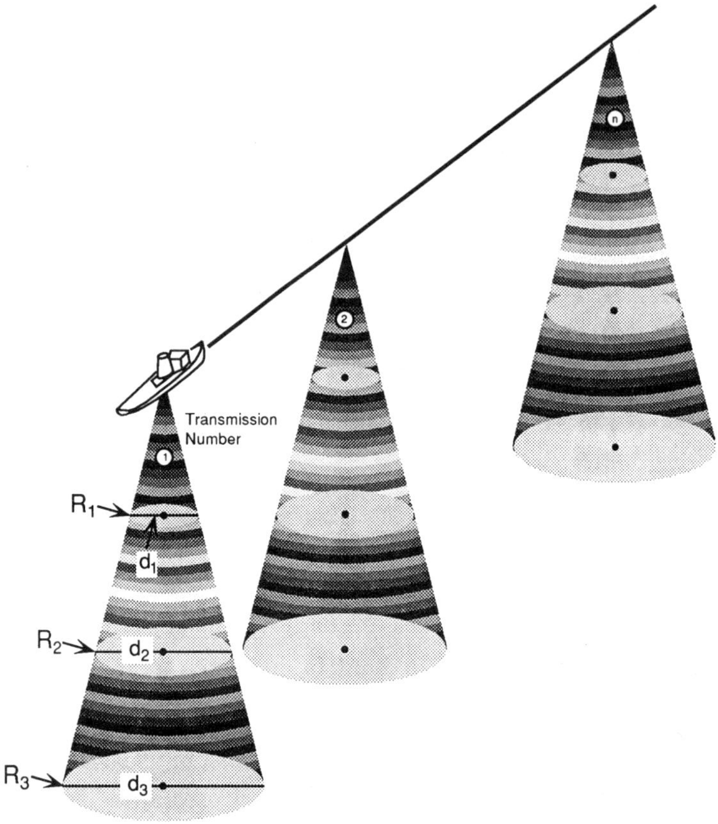

Figure 3-1. Volume sampled during successive transmissions ("pings") by an echosounder as the survey vessel proceeds along a transect. The area sampled increases as the square of range (R) from the transducer.

ferent species of fish, however, so that target identification with trawls, purse seines, etc., is an integral part of any acoustic survey.

The equipment used in acoustic surveys can be relatively simple in single species situations (trout or sockeye salmon in oligotrophic lakes, for example), where the fish are sufficiently dispersed that separate individuals can be identified by their returning echoes and counted. More complex equipment must be employed in situations where several species are present in the survey area and echoes from several individuals return to the survey vessel simultaneously.

In the latter case the returning echoes must be summed by a process known as echo integration, and frequent net sampling is required to verify that the echoes are being generated by the target species being surveyed. Carefully calibrated acoustic and echo processing equipment is required for useful echo integration results, but can open up the possibility of obtaining real-time estimates of abundance and detailed mapping of distributions.

ACOUSTIC SYSTEMS

Sound Propagation

Most acoustic surveys of abundance are carried out with sonar systems that direct their energy directly downward, along an acoustic axis that represents the shortest distance from the survey vessel to the bottom. A single transducer serves both to propagate the acoustic energy and receive the returning echoes it creates. This type of design is referred to as an echosounder, and is commonly employed on commercial fishing vessels around the world. Other designs of sonar systems are occasionally employed in abundance surveys (see Chapter 5 for an example) but the echosounder has the advantage of minimizing the refraction (bending) of the acoustic beam by discontinuities in the water column created by temperature and salinity gradients (Fig. 3-2).

Transducers come in a variety of designs and materials, and serve to convert the electrical energy produced by a transmitter into a series of pressure waves at the transducer face. The transducer oscillates and physically moves the adjacent water particles so that pressure waves are induced (Fig. 3-3). The transmitting system and transducer are designed so that these pressure waves

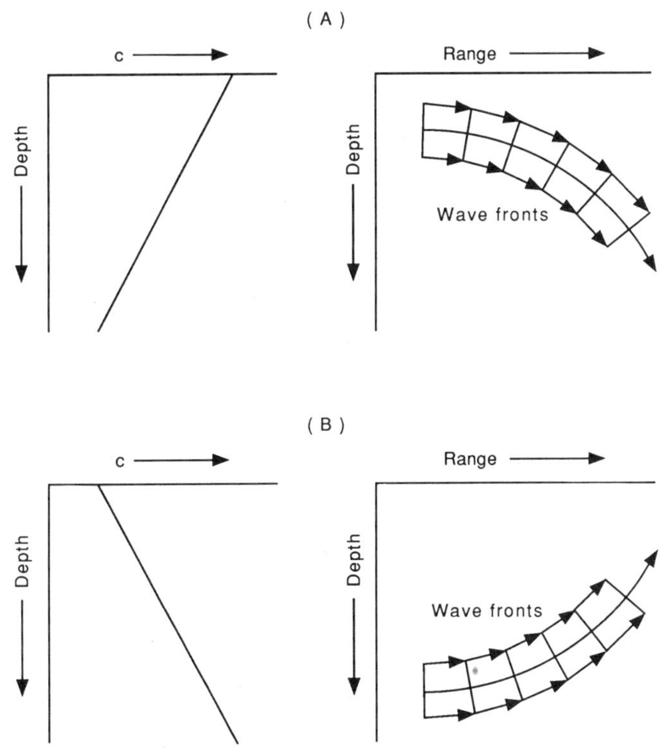

Figure 3-2. Refraction of an acoustic beam when the speed of sound (c) decreases (A) or increases (B) with depth.

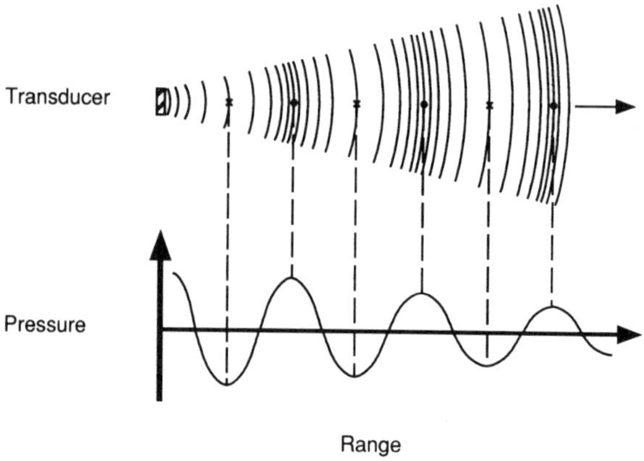

Figure 3-3. Propagation of an acoustic wave produced by the vibration of a transducer.

are propagated at a fixed intensity, characteristic operating frequency, and for a carefully controlled duration of time. The wave frequency (f) corresponds to a specific wavelength (λ) according to the relationship:

$$c = f\lambda$$

where c = the speed of sound in water (m/sec)
f = frequency in cycles per second or hertz (Hz)
λ = wavelength (m)

The operating frequency has an important bearing on the attenuation (absorption) loss that occurs as the sound wave travels through the water column (Fig. 3-4), and is a major determinant of the distance over which a wave can be propagated. The operating frequency can eventually be filtered out by the signal processing equipment (Fig. 3-5), since it is the energy contained in returning echoes rather than the carrier frequency that is of interest.

It is the transmission duration of the individual pulse (pulse duration in seconds, or τ) and the corresponding length of the pulse (in meters) that ultimately determines our ability to resolve fish targets in range. If the speed of sound (c) is 1500 m/sec, for example, a pulse duration of 1 millisecond (ms) will produce a pulse that is 1.5 m in length.

Echoes from the leading edge of this pulse will return toward the echosounder during time $c\tau/2$, combine with those generated by the trailing edge of the pulse, and the combined echoes will arrive at the echosounder simultaneously (Fig. 3-6). At any given instant of time then, the echosounder will be receiving echoes from a segment of the water column that is $c\tau/2$ m thick (0.75 m in our example). If two fish are closer in range than $c\tau/2$, their returning echoes will overlap. The shorter the pulse duration and pulse length, the greater our ability to resolve single fish targets, or discriminate fish targets from the bottom signal.

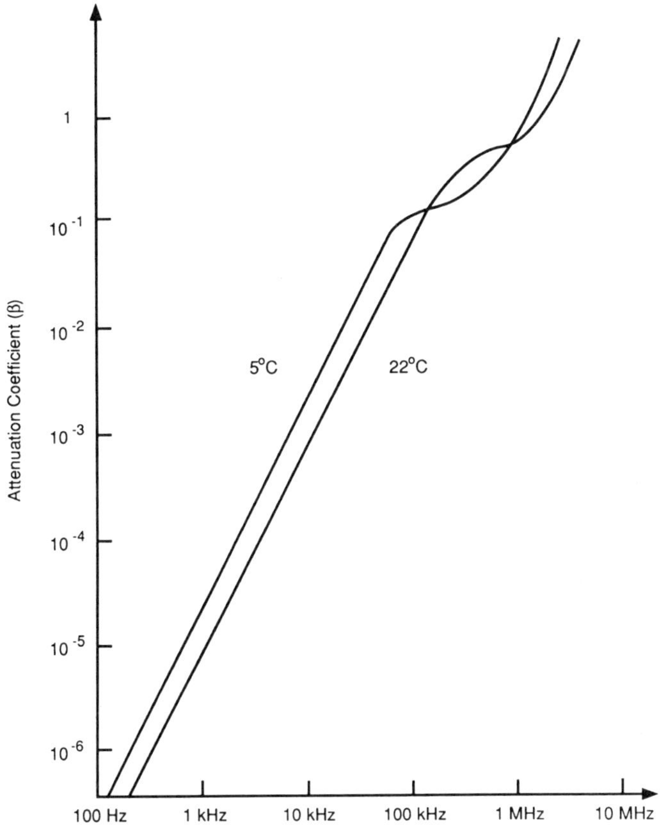

Figure 3-4. Exponent for rate of acoustic energy loss per meter (β) as a function of operating frequency of the transducer and ambient temperature. After Forbes and Nakken (1972).

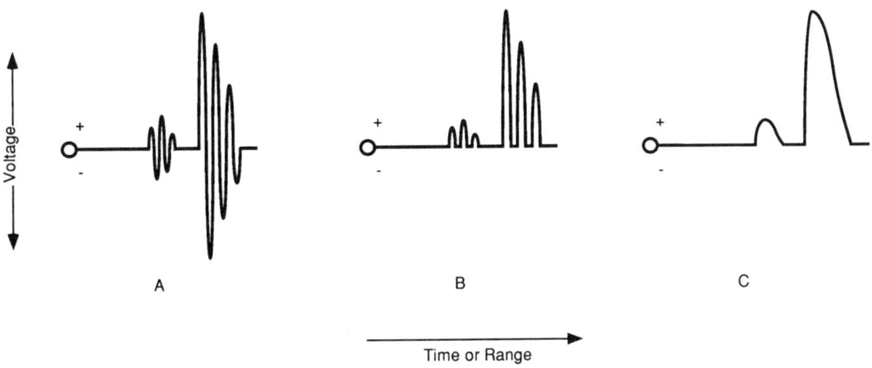

Figure 3-5. Effect of passing a sonar signal through processing filters. (A) Input signal, (B) removal of negative polarity portion of signals (rectification), (C) same as (B), but high (carrier) frequency oscillations removed.

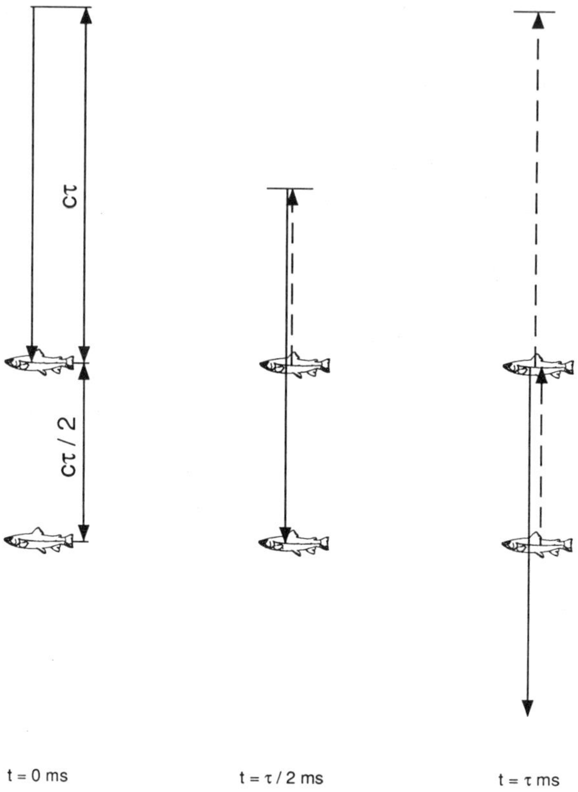

Figure 3-6. At any instant in time, an echosounder is receiving echoes from a segment of the water column that is $c\tau/2$ m thick. The solid lines show the progress of the transmission pulse, while the dashed lines show the progress of echo returns back toward the transducer.

Several pulses are generated per minute by the transducer. This same transducer must also function in a receiving mode (transforming the pressure waves from returning echoes into electrical energy), however, so that a pause is required between transmissions. This pause must be at least long enough to allow for the time required for the sound generated by the transducer to traverse the entire water column and return. A longer pause is necessary to avoid receiving "second echoes" from a given transmission, generated by energy that has made a full round trip from the transducer to the bottom, then been reflected toward the bottom once again. The pulse repetition rate will consequently be determined by bottom depth, ranging from about 4 transmissions ("pings") per second in shallow water to about 20 transmissions per minute in 1000 m.

The pressure (p) generated by the transducer is usually measured in micropascals, μPa, where a pascal is one newton per square meter. As a pressure wave propagates, a given amount of energy per second (watts) will flow across a unit area positioned normal to the direction of wave propagation, producing an intensity I (watts per square meter, W/m^2) that is proportional to the square of pressure. The exact relation is

$$I = p^2/\rho c$$

where ρ is the density of the water (kg/m^3) and c is the speed of sound in water.

The pressure wave generated by the transducer spreads as the square of the distance it travels through the water column, giving rise to a sample volume that can be crudely portrayed by the cones shown in Figure 3-1. The angle subtended by the main acoustic beam ("beam angle" or "beam width") is a function of the operating frequency and physical size of the transducer, and can be designed to be as wide or narrow as desired. The "beam angle" (θ) is defined as the angle at which the intensity has decreased to half the value in the center of the beam (on the acoustic axis), and can be approximated by

$$\text{Sin } \theta = 0.44\ \lambda/l$$

where θ = half-angle beam width
λ = wavelength (m)
l = length of transducer face (m)

From this it is clear that larger transducers and higher operating frequencies (for example 120 versus 38 kHz) will result in narrower beams. Transducers can be designed so that the fore-aft and port-starboard beam angles differ, but these will be the same for a circular transducer.

Tradeoffs are involved here, and at first glance it may appear that the beam should be as wide as possible to maximize the area being sampled. However, since echoes from fish at the same range all return to the transducer simultaneously (Fig. 3-7A), the ability to distinguish individual fish targets (of critical importance when echo counting) is enhanced by employing a narrow beam. Also, for a wider beam, the noise level will be higher relative to the returned signal.

In addition, the seabed or lake bottom is always a very strong reflector of acoustic energy, and once any portion of the beam contacts it, its echo is so strong that it dominates the subsequent signal returning to the transducer. This gives rise to a near-bottom "dead zone" (Fig. 3-7B), which increases in size with rolling and pitching of the transducer, beam angle, pulse length, bottom roughness, and range from the transducer. On flat bottoms at moderate depths (less than 100 m), the dead zone will be on the order of 0.5 m with commonly used equipment (Aglen 1989). Along steep slopes in deeper water, the dead zone may exceed 10 m. As a result, it is important to use the narrowest beam angle and shortest pulse length possible when surveying species that are found near the bottom.

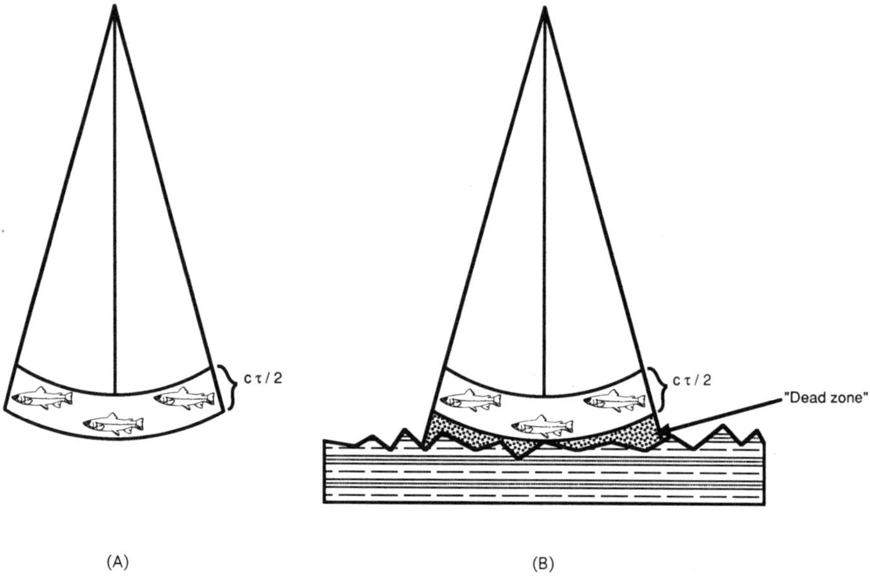

Figure 3-7. Discrimination of echo returns from fish targets. (A) Echoes from fish located at the same range will return to the transducer simultaneously and be indistinguishable. (B) The bottom echo completely masks the presence of fish in the "dead zone," sharply reducing the effective sampling volume. In addition, the echoes from all three fish will overlap with the bottom signal, making them almost impossible to discriminate.

In fact, the volume sampled is far more complex than portrayed by the "beam width" or the simple cones in Figure 3-1, since

1. The pressure wave generated by the transducer varies widely in intensity over the cross section of the beam (Fig. 3-8), rather than being uniform throughout.
2. A certain amount of background acoustic "noise" and reverberation (echo returns from nonfish targets such as plankton, suspended particles, gas bubbles, etc.) exist.
3. The amount of energy per unit area decreases as the square of range (R, or distance from the transducer) due to spreading.
4. Additional acoustic energy is lost at an exponential rate ($e^{-\beta R}$), through absorption and conversion into heat (attenuation).
5. When the level of energy per unit area within an echo returning from a fish falls below that generated by background noise and echoes from nonfish targets, we will no longer be able to detect its presence.

Given adequate technical support, most of these factors can be controlled and quantified in such a way as to accurately define the volume sampled per transmission. Nevertheless, the probability of detecting a given fish will de-

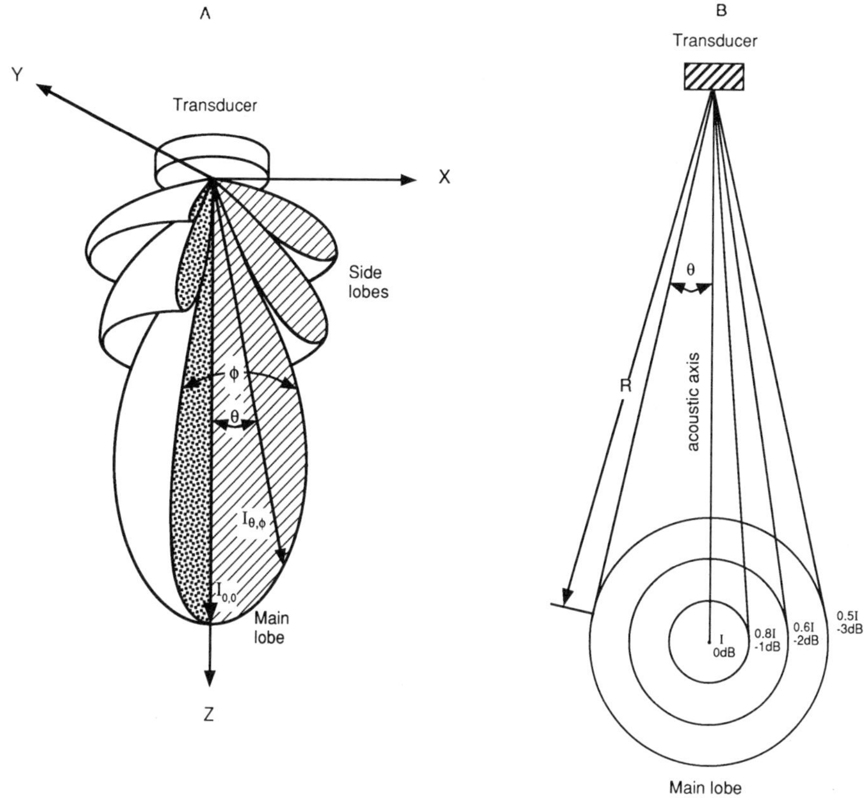

Figure 3-8. Directivity pattern of an acoustic beam. (A) Variation in the intensity (watts/m^2) of the energy field produced by a circular transducer at constant range, but different angles (θ,ϕ) off the acoustic axis. $I_{0,0}$ = intensity on the acoustic axis. (B) Cross-section of the main lobe, showing the loss in acoustic energy with increasing beam angle (θ). R = range from transducer.

cline with increasing angular distance off the acoustic axis (Fig. 3-8), and with range from the transducer.

The variability in sound intensity over the cross section of the acoustic beam (beam pattern, or directivity index) is dependent on the engineering of the transducer. A transducer is made up of several independent elements configured so that these multiple point sources interact to produce a well-defined beam, the energy being concentrated around the acoustic axis with little going into side lobes (Fig. 3-8). For scientific work, it is important that side lobes be minimized, since these can pick up unwanted noise and reverberation. Mitson (1983) and MacLennan and Simmonds (1992) can be referred to for detailed accounts of the design and construction of acoustic transducers.

Noise levels in the echosounder system can also be reduced through proper system engineering. Noise differs from reverberation in that it will be received even when no acoustic energy is being generated by the transducer. Background noise is generated by a variety of sources, over a wide variety of frequencies (Urick 1975) including:

1. ship noise due to machinery, interference from electrical equipment, water flow over the hull and propeller cavitation;
2. "self"-noise induced in the echosounder's amplifier and processing circuitry;
3. breaking waves, tides, oceanic turbulence, rain and wind; and
4. intermittent biological noise caused by fish, crustaceans, and marine mammals.

While many of the noise sources in the first two categories can be minimized by ship and electronic design (obviously a major focus of research and engineering in submarine-antisubmarine warfare), noise can never be eliminated. Consequently, a second requirement of the transducer is that it transmit over as narrow a range of frequencies (bandwidth) as possible, since the greater the bandwidth produced by the transducer, the greater the amount of noise that will eventually be received by the system. The triggering and turning off of the transducer results in the generation of frequencies other than the main operating frequency (f), and if the pulse duration is τ seconds, then frequencies are generated at intervals of $1/\tau$ Hz on each side of f (Mitson 1983). We have already seen that range resolution can be improved by employing shorter pulse lengths. For any given frequency, however, bandwidth must be increased as pulse length is reduced, and an unfortunate side-effect is that the amount of noise increases as a by-product.

Echo Reception

At any given range (R) from the transducer, the sound intensity on the acoustic axis will decline with spreading and attenuation, and is given by

$$I = \frac{I_0 e^{-\beta R}}{R^2}$$

where I = intensity (W/m^2)
I_0 = source level of intensity produced by the transducer, as measured at a range of 1 m on the acoustic axis
β = attenuation coefficient

Since the level of sound transmitted varies over the cross section of the beam, the level of intensity at any bearing off the acoustic axis will be less than that on the acoustic axis (Fig. 3-8). The sound intensity at any given range from the transducer and bearing off the acoustic axis (θ, ϕ) will be given by

$$I = \frac{I_0 e^{-\beta R}\, b(\theta, \phi)}{R^2}$$

where $b(\theta, \phi)$ is a function that describes the beam pattern relative to the acoustic axis. This has a value of 1.0 on the acoustic axis, and declines as θ increases.

When the sound strikes a fish, only a fraction of it will be reflected back toward the transducer. For a fish on the acoustic axis, this can be expressed as

$$I_r = kI$$

where I_r = reflected sound intensity (W/m^2), measured at a range of 1 m from the target
I = incident sound intensity
k = a constant dependent on the reflective properties of the fish

A common way of expressing k is in terms of "acoustic cross section" or "equivalent cross section" (σ). In this convention $k = \sigma/4\pi$, and a perfectly reflecting sphere with a radius of 2 m would have a k value of 1.0.

Target strength is highest when the incident sound encounters an object that is substantially different from water in density and sound transmission characteristics, and this accounts for the extremely strong reflective properties of the bottom substrate and the air–water interface at the surface. In fact, the surface and bottom are such good acoustic reflectors that it is often possible to receive "false bottom" signals that originate from a second round-trip by the acoustic energy originating at the transducer.

For many fish targets, the gas in the swim bladder is responsible for 90–95% of the total energy reflected (Foote 1980). This has important ramifications in survey design, since the size of the swim bladder can vary substantially when fish undertake vertical migrations (Ona 1990). Furthermore, fish that lack a swim bladder (e.g., dogfish, *Squalus acanthius*; Atlantic mackerel, *Scomber scombrus*) are usually poor candidates for acoustic surveys, since they are difficult to detect unless aggregated.

For a fish located off the acoustic axis, the strength of the returning echo will once again depend on bearing off the acoustic axis and the beam pattern. Sound traveling back toward the transducer will once again be reduced in intensity by spreading, attenuation and the beam pattern effect, so that the level reaching the transducer will be given by

$$I = \frac{kI_0 e^{-2\beta R} b^2(\theta, \phi)}{R^4} \tag{3.1}$$

Notice that this equation contains two-way corrections for spreading, attenuation, and beam pattern.

So far, the discussion has been confined to situations where a single fish target is present in the volume being sampled during any given instant. This volume (V) will be $c\tau/2$ m thick and have a surface area determined by the beam pattern and range from the transducer. Ehrenberg (1979) defines this as

$$V = R^2 \pi c\tau \int_0^{2\pi} \int_0^{\pi/2} b^2(\theta, \phi) \frac{\sin(\theta)}{2\pi} \, d\theta \, d\phi$$

or

$$V = R^2 \pi c\tau \int_0^{\pi/2} b^2(\theta) \sin\theta \, d\theta$$

for a circular transducer.

Many authors (Burczynski 1979, Johannesson and Mitson 1983) define V in terms of the "elemental solid angle" ($d\Omega$, in steradians), as shown in Figure 3-9. The sampling volume can then be defined as

$$V = R^2 \frac{c\tau}{2} \int_{\Omega=0}^{4\pi} b^2(\theta, \phi) \, d\Omega, \qquad \text{or } R^2 \frac{c\tau}{2} \psi$$

where ψ may be regarded as the solid angle subtended by an "ideal" beam pattern, with $b^2(\theta, \phi)$ being uniform for all angles less than ψ, and 0.0 for all angles greater.

In any event, it should be clear that the directivity pattern of the transducer must be known, and integrated over in an appropriate manner if the exact sampling volume is required. This is the case in many survey situations, where the distribution of the target species is such that several individuals occupy the volume being sampled (Fig. 3-7). Their echoes return to the transducer simultaneously, and single targets cannot be discriminated. Assuming that the fish are spread uniformly throughout the acoustic beam, and the volume being sampled at any instant is as shown in Figure 3-9, the echoes from a density of D_v individuals/m^3 will sum such that their combined echo strength will be

$$I = \frac{D_v(c\tau/2)\psi \, kI_0 e^{-2\beta R}}{R^2} \tag{3.2}$$

As range increases, the increase in the volume sampled by the acoustic beam is exactly offset by the reduction in energy per unit area, so that an R^2 term in the numerator has canceled out a similar term in the denominator.

The pressure wave returning to the transducer has an energy level of I watts/m^2 and must now be converted to electrical energy (watts).

From basic physics it will be recalled that

$$W = V^2/r$$

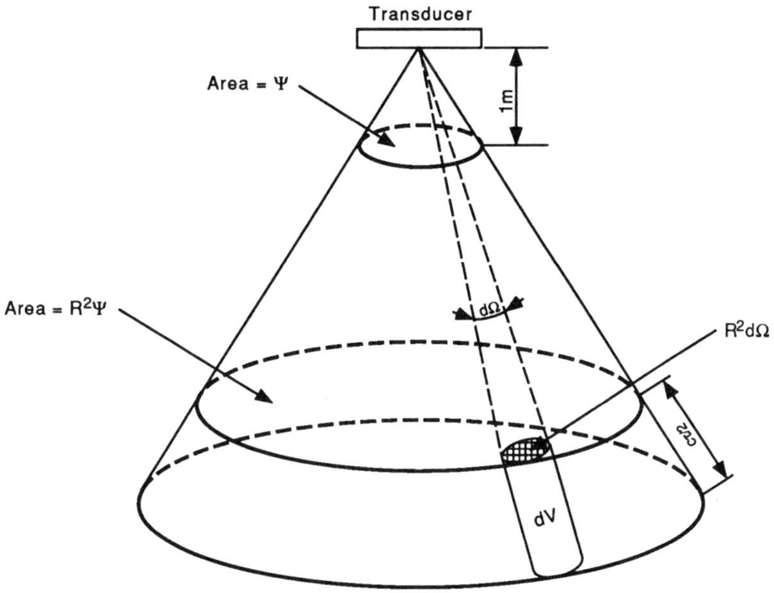

Figure 3-9. At any range R, the acoustic sampling volume is defined by the area $R^2\psi$ (obtained by integrating over $d\Omega$, the elemental solid angle), and thickness $c\tau/2$.

where W = electrical power (watts)
V = voltage
r = resistance.

The efficiency of converting acoustic energy to electrical energy in an echosounder system is relatively constant in well-designed systems, and can be adjusted for by careful calibration. Echo intensity is therefore proportional to V^2, and the voltages generated by the transducer can be digitized or amplified in a variety of ways to attain the objectives of the user.

Signal Processing and Display

The voltage signals produced by the transducer are relatively weak, and must first be amplified before they are ready for further display. The amplifier is tuned to the specific operating frequency of the transducer, and responds only to a limited bandwidth. On most echosounders, the level of amplification can be controlled by a manual gain control, which can be used to turn up the gain to a point just below that where the background noise begins to become evident on the display.

An important requirement is that output voltage from the amplifier responds in a linear fashion to input voltage, at least between the extremes of signal voltage likely to be encountered during the survey. Saturation (where voltage response is no longer linear) occurs in many systems, and the difference between the minimum usable signal at the receiver input and the maximum input signal that does not cause saturation defines the "dynamic range" of the amplifier.

Noise is processed in the same way as the signals, and can never be eliminated, although certain types of thresholding or correlation techniques can be used to reduce its impact on survey results. For example, many sources of noise (e.g., sea noise) are independent of depth and can be filtered out with a simple nonlinear (to adjust for the TVG correction) threshold gate. The amplifier itself can generate noise internally, and a poorly designed amplifier can severely compromise the results from a survey (Aglen 1989).

Once amplified, the signals are still in the form of a pulse comprising a certain number of cycles at the echo sounder (carrier) frequency. For display or integration purposes, the signal is often demodulated and "rectified" (Fig. 3-5). This process can be carried out either electronically or digitally, and can remove the carrier frequency and negative half of the signal, resulting in a unidirectional DC waveform.

Perhaps the simplest display is the oscilloscope or cathode ray tube (CRT, Fig. 3-10), commonly used in echo counting. As can be seen in Figure 3-5, the voltages received from each transmission are portrayed in terms of time, and can readily be converted to voltage versus depth (range from echo sounder) knowing the speed of sound (m/sec). Because of energy dissipation due to spreading and losses due to attenuation, the echo return from a fish located deep in the water column will be much weaker than for an identical fish in shallow water. As a result, signals are usually systematically amplified over time (time-varied gain or TVG) to adjust for these effects. A "20 log R" TVG adjusts for one-way losses due to spreading and two-way losses due to

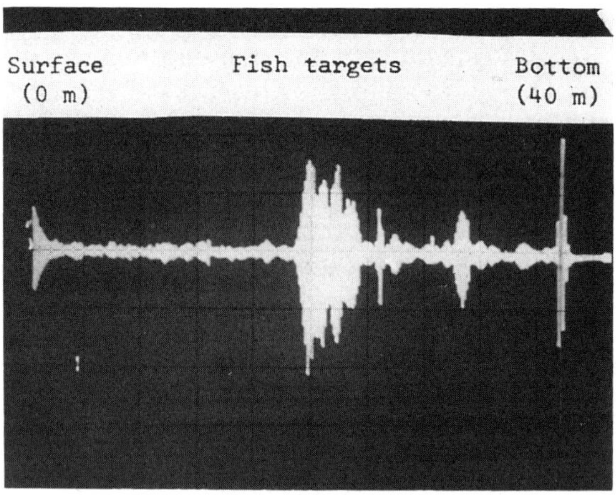

Figure 3-10. Oscilloscope trace of a single transmission "ping" in a freshwater lake. Voltage is presented as a function of time (t). Range (R, in meters) is determined from the relation $R = ct/2$. Courtesy of R.E. Thorne, BioSonics, Inc.

attenuation, increasing the gain in proportion to $R^2e^{2\beta R}$, while a "40 log R" TVG gives a two-way adjustment of $R^4e^{2\beta R}$.

As range increases, the increase in the volume sampled by the acoustic beam is exactly offset by the reduction in energy per unit area, so that a 20 log R TVG (one way correction for spreading of returning echoes) is all that is required by commercial fishermen interested only in comparing total density at different depths in the water column. With this TVG correction, a density of 500 fish/m^3 at 100 m depth will register the same electrical energy (wattage) as the same density at 30 m. A 20 log R TVG is also used when integrating multiple echoes (see Equation 3.2).

When counting echoes from individual fish, however, it is critical that the returns from different depths register the same voltage, so that we can be sure that we are counting single targets of the appropriate strength. In this case 40 log R is the appropriate TVG. However, differences in individual echo strength will still exist due to variations in the location within the beam pattern, or orientation of the swim bladder with respect to the transducer, even when the fish being observed are identical. Nevertheless, the voltage from identical fish (same target strength) at the same angle off-axis will be the same, independent of range.

A wide variety of other display media are available to the user, including paper echograms (Figs. 1-4 and 3-11), CRT displays that use technology from television (Fig. 3-12), and speakers that convert the signal into audible sounds. Each of these can be useful as aids in identifying the species contributing to the returning signal, and determining where to deploy sampling nets for precise target identification. In many instances, several alternate display devices will be used simultaneously.

Most display units used for survey work will be equipped with features that allow scale expansion for depths selected by the user, and "bottom locking" (Fig. 3-11) so that echoes from the near-bottom region can be stored in a memory and played back after the bottom signal has been received. This latter feature allows continuous expansion of the near-bottom region without frequent readjustment of the expansion range.

A "whiteline" or "grayline" feature is usually available (Fig. 1-4), and is generated by stopping the display of all signals for a brief instant following reception of the bottom signal. This serves as an aid in discriminating near-bottom concentrations of fish from the bottom.

Color (CRT or echogram) displays convert signals of different intensities to a specific colors for display. The colors can often be adjusted to the preference of the user. In Figure 3-11, for example, the strongest signals corresponded to red and the weakest to violet in the original echogram.

While many fishermen have become quite sophisticated in their ability to interpret these displays and guess which species are responsible for a given echo trace, this is usually based on making several fishing sets in a given season, depth, and geographic area, together with an intimate acquaintance with the characteristics of the display equipment at their disposal. The ap-

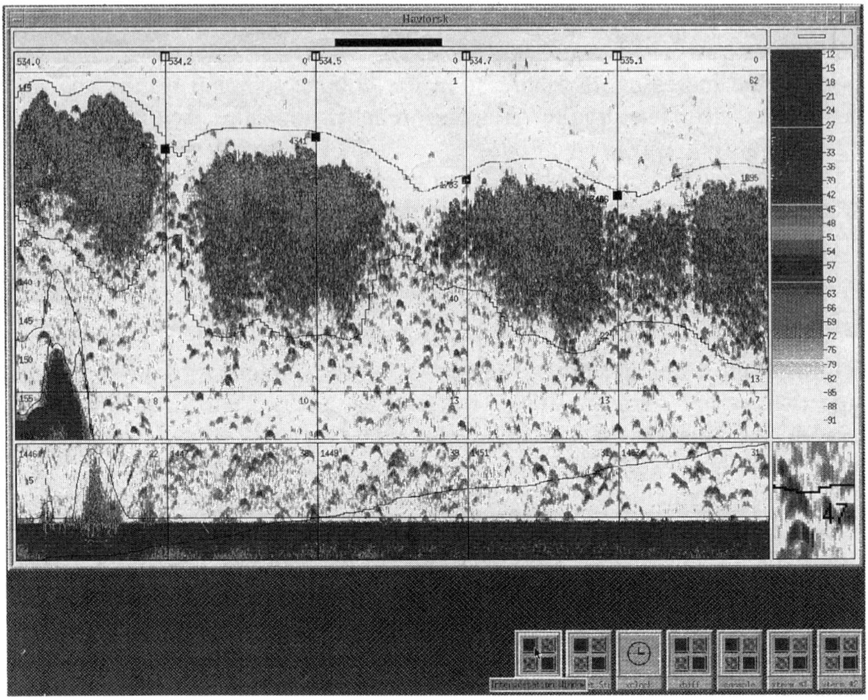

Figure 3-11. Reproduction of a color echogram display from the Bergen Echo Integrator. In the original echogram, each color (chosen from the "palette" at right) represented a different range of signal intensities. The bottom (far left, middle of display) is denoted by the dark (red in original) trace. A "bottom locked" expansion of the 10 m closest to the bottom is shown immediately below the full echogram. The operator has delineated the desired integration bounds by the lines and solid squares indicated on the echogram. Courtesy of H.P. Knudsen, Norwegian Institute of Marine Science.

pearance of the echo trace from a given school of fish will depend on vessel speed, paper speed (scan rate for color CRT displays), and gain settings, all of which can alter the size, shape, and apparent density of a school. Considerable experience with an echosounder system and fishing experience in an area are required before competent judgments can be made as to the identification of the species contributing to a given echo return.

An additional source of information on the size and species composition of the fish generating echoes lies in the target strength or length frequency displays available with some systems (Fig. 3-12). While these can be an important aid in target identification, they are only as good as the equipment on which they are based. Target strength distributions based on dual beam or split beam transducers (see Estimating Target Strength) can be quite useful in target identification, since the target's bearing off the acoustic axis can be determined with these systems. Even with these systems, however, it is im-

Figure 3-12. Video echosounder display. Estimated length composition of the fish targets is presented in the histogram at lower right. Courtesy of SIMRAD A/S.

perative that an adequate sample of single targets is being isolated, and that the "aspect" (orientation of the swim bladder) presented to the transducer is relatively constant. Even if only two species of substantially different size are present in the water column, their target strength distributions can show considerable overlap due to individual variations in size and aspect. Rose and Leggett (1988) found that they were unable to classify schools of cod, capelin, and mackerel on the basis of mean target strength alone (Fig. 3-22).

Conversion of target strength data to data on length composition is even more problematic, since it is dependent on empirical equations that may fail to take into account important sources of variation (notably swim bladder morphology and dynamics) between and within species. As such, the length frequency displays available on some systems serve to provide only a rough idea of the actual length composition of the targets.

Storage of the data is an important part of any survey, since a permanent record is usually desired. This can be used to process the data in greater detail once ashore. Echograms and cassette recordings from color CRT displays form one part of this record, but it is also important that the actual echo voltage record be permanently stored. This can be done by storing the analog signal directly, using audio or video tape. Alternatively, the signals can be digitized or integrated at convenient intervals, then stored on a computer disk file or optical disk. The method used to store the data (and data backups) will vary from application to application, depending on the facilities and financial resources available.

TARGET IDENTIFICATION

Since the information from the echosounder displays available to the user are insufficient in themselves to be certain as to the size and species composition of the fish (or other organisms) contributing to echo returns, alternate means of target identification are an important part of acoustic surveys.

Species Identification

One approach to species identification that has been successful in experimental situations has been to quantify a number of parameters such as position of schools in the water column, target strength distributions, school density, school dimensions (length, width and depth), distance between layers within the school, etc., from echo return data, and use discriminate analysis to determine the species composition of the targets. This represents the introduction of quantitative methodology in interpreting the species composition of acoustic targets, making full use of the information contained in the echo returns. As such it is a logical extension of the intuitive approach used by scientists and commercial fishermen in the interpretation of the information available from an echogram. In one experimental study (Rose and Legget 1988), it was possible to classify 93% of the cod, capelin, and mackerel schools observed correctly using a discriminant function with 5 variables. The authors caution, however, that the form, size, and density of fish schools are likely to vary with time and environmental conditions and the discriminant functions would consequently require frequent recalibration.

Although supplementary acoustic technology has been developed for target identification work, it has not yet been used in routine acoustic surveys. One technique involves the transmission of multiple frequencies until one is found that causes the swim bladder to resonate (Holliday 1977a). At resonance, the energy reflected by a given fish can increase by about a factor of 5 (Forbes and Nakken 1972). The resonant frequency (usually less than 10 kHz for fish from 20 to 90 cm) can then be related to swim bladder size and fish length. One problem with this technique is that swim bladder volume (and consequently resonant frequency) can change during bathymetric migrations, feeding etc. (see Target Strength).

The swim bladder is not the only part of the body that reflects sound, however (Nash et al. 1987), and target strength changes substantially at different frequencies. Recently, wideband echosounders have been developed that transmit over a broad range of frequencies, and preliminary work has shown promising results in separating different species in experimental cages. Figure 3-13 (Simmonds and Armstrong 1990) shows one set of results from such experiments indicating that it is possible to distinguish herring from mackerel (and both of these from cod and saithe) on the basis of the acoustic "signatures" generated by changes in target strength at different frequencies.

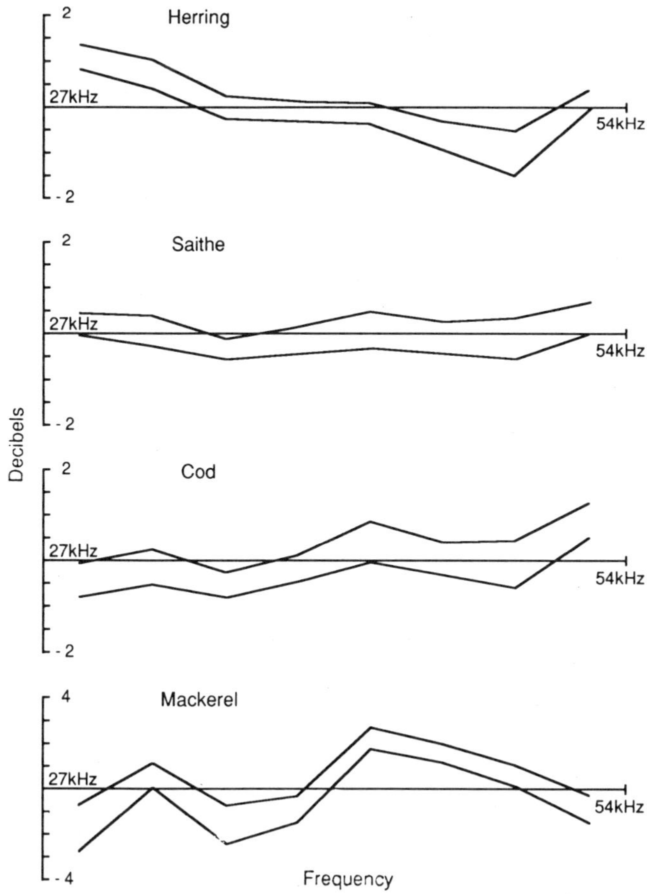

Figure 3-13. Average frequency response spectra ±1 SD for four species of fish. From Simmonds and Armstrong (1990).

The cod and saithe were more difficult to differentiate, but could be discriminated with 90% accuracy given 1000 transmissions.

A second acoustic approach (Holliday 1977b) has been to measure the Doppler shift caused by the tail of a swimming fish, and relate this to tail beat frequency and fish length. As is the case with wideband echosounders, however, this technique has yet to be used during routine surveys, evidently because of the additional equipment required to use it.

A wide variety of camera techniques have been used in target identification, including underwater television and free-falling "drop" cameras that take flash photographs at intermittent intervals during their descent. However, these have yet to overcome the technological problems involved in deploying them routinely from survey vessels, or the avoidance problems associated with their use. Many fish see well at low levels of light, and have detected the camera (or the noise associated with its deployment) and reacted before they can be observed.

While additional remote techniques that supplement our current capabilities in species identification may eventually be developed, it is only by direct sampling that information on sex or age composition can be obtained. Consequently net sampling techniques that provide an unbiased sample of the fish contributing to echo returns will always be an integral part of acoustic surveys.

While purse seines are useful in this regard when the target species is a schooling pelagic fish (herring, anchovy, sardine, etc.), they must be used near the surface, where fish are most sensitive to the light, noise, etc., generated by the survey vessel. As such, their efficiency can vary widely with ambient conditions. Misund (1990) found that the capture success of purse seines ranged from 90% during the nighttime winter fishery for Norwegian spring-spawning herring to 64% during the daylight summer fishery for North Sea herring.

Midwater trawling is the method most commonly used to determine species composition in marine surveys. Since catches from a bottom trawl are often dominated by organisms that inhabit the lower 1 m of the water column, where acoustic techniques are usually ineffective, they provide information that is of limited use during acoustic surveys.

The ability of fish to avoid survey vessels and bottom trawls has been discussed in Chapters 1 and 2. Since fish in midwater have an additional direction (downward) in which to escape from an oncoming trawl, trawl avoidance is an even more difficult problem there. Commercial fishermen have developed a number of techniques to overcome this, the most notable of which are the development of trawls with an extremely large opening (Fig. 3-14), and towing them at high speeds. Commercial midwater trawls used for pollock in the eastern Bering Sea frequently have a vertical mouth opening of 60 m and a horizontal opening of 100 m. The wings of these trawls are made up of extremely large meshes (as much as 25 m) or sets of parallel ropes to reduce drag and allow the net to be towed at speeds of 5–6 knots despite their size.

Midwater trawling is typically aimed at specific concentrations that have been identified by the echosounder, and the trawls are equipped with transducers that monitor their position in the water column (Fig. 3-15). These transducers transmit continuous information on the location of the trawl and the quantity of fish entering the net to the survey vessel,where it is displayed on special chart recorders (Fig. 3-16) or CRT displays. Deploying these trawls takes considerable time, and it is seldom possible to make more than a few sets per day and still accomplish all of the goals of an acoustic survey. Consequently, many aggregations that are detected on the echosounder must be classified on the basis of previous trawl sets, and past experience in the area. It is seldom possible to make enough trawl sets to be certain that all aggregations encountered have been accurately identified as to species.

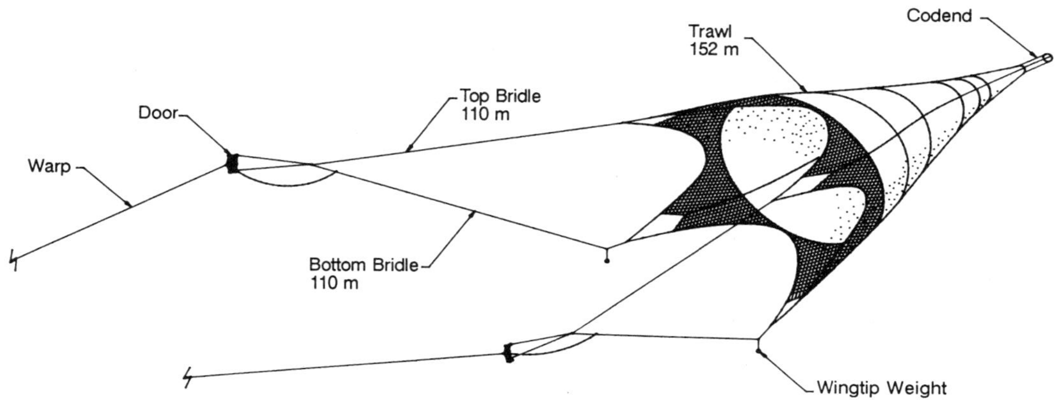

Figure 3-14. A midwater trawl designed for commercial fishing, with large-mesh (12 m) wings, and wingtip weights used to maintain vertical spread. Mouth dimensions (at wing ends) when fishing are 73 m (horizontal) and 49 m (vertical). Courtesy NET Systems Inc.

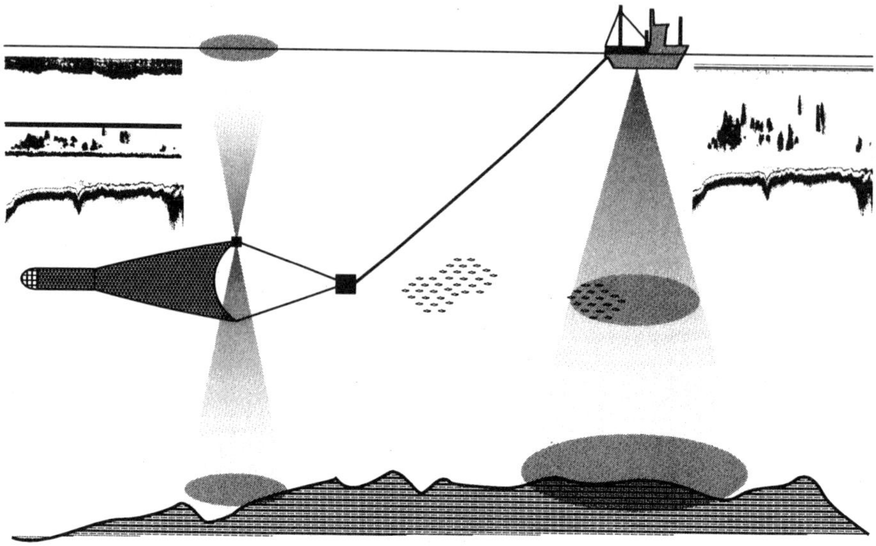

Figure 3-15. Aimed midwater trawling. Information on school density and depth from the ship's echosounder (far right) is used to aim the trawl, while information from the transducer on the net (display at far left) is used to monitor its position in the water column.

Furthermore, the large mesh sizes used in the wings together with the avoidance behavior employed by midwater fish make it difficult to be certain that the species and size composition of the fish in the catch are truly representative of those contributing to the echo returns.

Nakashima (1990) evaluated the selectivity of a Diamond IX midwater trawl by attaching special sampling bags to the outside of the net, and comparing the catches in these with those in the cod end. Both the sampling bags (4-m^2 opening) and the liner for the cod end were made of 6.3-mm nylon mesh. Nakashima found that small capelin (*Mallotus villosus*), sand lance (*Ammodytes dubius*), arctic cod (*Boreogadus saida*), and juvenile groundfish escaped from the intermediate portions of the trawl in significant numbers, with no arctic cod being represented in the cod end. This resulted in overestimation of the biomass and mean size of capelin during acoustic surveys for this species off eastern Canada.

As a consequence, acoustic surveys are most successful when used in situations where the fish aggregations encountered are dominated by a single species with a narrow size range. In many cases, this can be facilitated by defining several area–depth–time strata such that a single species and size group dominates each stratum.

In multispecies situations, a large number of trawl hauls is required to properly identify the species composition of the aggregations in a given stratum, and when the catches from these sets contain more than a single species (haddock and cod in the Barents Sea or herring and Pacific hake in the

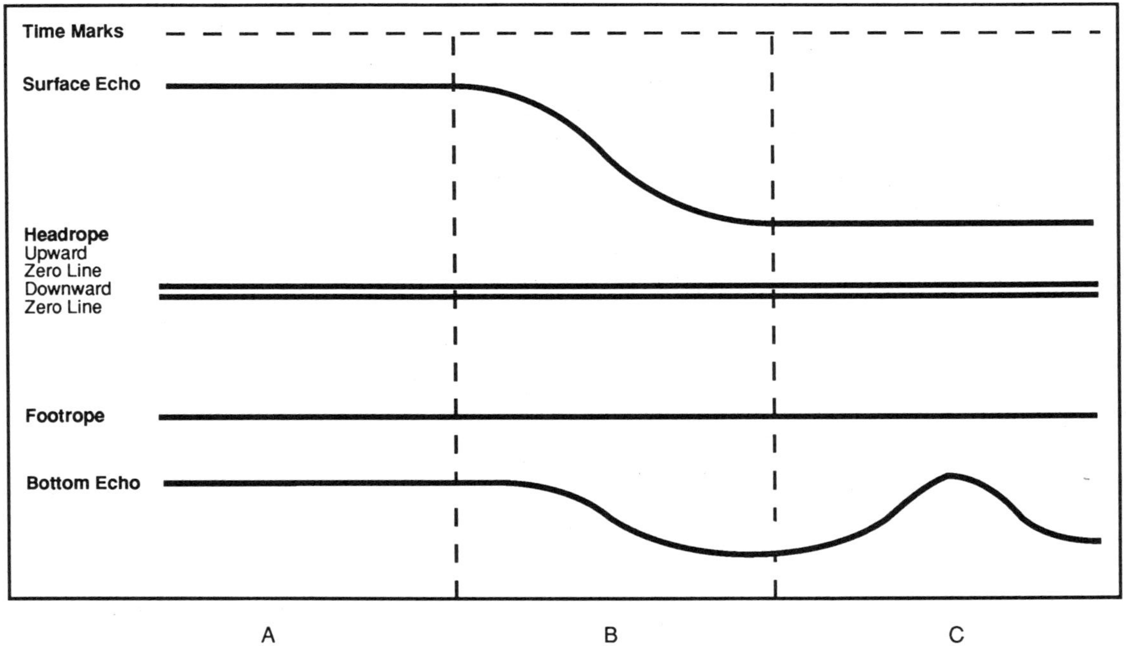

Figure 3-16. Information provided to a wheel-house chart recorder by the net transducer. (A) Headrope to surface distance constant, footrope to bottom distance constant. Net at a steady depth and bottom flat. (B) Headrope approaching surface echo, and moving away from bottom. Net moving up in water column. (C) Headrope to surface distance constant, but bottom echo approaches footrope then falls away. Net at steady location in the water column, but passing over a rock pinnacle.

northeast Pacific, for example), allocation of the echo return data to different species becomes problematic. It has already been shown (Chapter 2) that trawl catches can be severely biased by avoidance of the survey vessel, trawl doors, and trawl, as well as selectivity by the trawl meshes themselves. As such, it should never be assumed that the size and species composition of these catches are representative of the aggregations detected by the echosounder without supporting evidence. Special behavioral and gear avoidance studies, or comparisons between independent estimates of abundance should be undertaken to show that this assumption is valid. Comparison of population estimates from surveys and catch-age analysis indicated that the use of species composition data from the trawl catches to allocate acoustic returns resulted in underestimates for cod and overestimates for haddock in the Barents Sea (Nakken and Ulltang 1983).

Surveys can often be timed so that the pelagic community is dominated by a single species, and target identification problems are relatively minor. In Cultus Lake, British Columbia, for example, acoustic surveys for juvenile sockeye salmon are done at night when the fish are feeding in the upper (mainly 20–25 m) portion of the water column (Nunnallee 1983). During the day, the fish are grouped closer to the bottom and are harder to quantify acoustically. A 3 × 6 m midwater trawl used to sample the aggregations being surveyed (Burczynski and Johnson 1986) showed that 95–96% of the fish captured were juvenile sockeye salmon. Due to the slow trawling speed (0.6 m/sec) employed and the relatively small size of the net, the midwater trawl could not be expected to catch larger fish (20–67 cm) that reside in the lake, such as kokanee (*Oncorhynchus nerka*), squawfish (*Ptychocheilus oregonensis*), and Dolly Varden trout (*Salvelinus malma*). However, previous gillnet sampling suggested that relatively small numbers of these species were in the lake during the acoustic surveys, and their impact on survey results could be disregarded.

Size and Age Composition

As was the case with bottom trawl surveys, the main parameters of interest are population size and/or biomass estimates for selected target species. A second goal of most acoustic surveys is to determine the length and/or age composition of these species, and this can be accomplished only by sampling with fishing gear and sampling techniques that provide representative samples of the population. Since midwater species are often able to detect and avoid the survey vessel and fishing gear, the assumption that survey catches are representative of the population should always be validated through special studies of avoidance behavior.

Midwater trawl sampling of acoustic targets is usually carried out in order to identify the species and size composition of specific aggregations, and random or systematic sampling is not possible. Midwater trawl sets are time consuming (1–3 hours per set), but must be taken with sufficient frequency to be confident that they are providing a representative sample of the length and age composition of the aggregations being censused acoustically. If size and age composition vary significantly between the hauls made within a given stratum, it is necessary to sample as many aggregations as possible, even when their species composition is known with certainty. Particular emphasis should be placed on adequate sampling of the largest aggregations detected.

The procedures used to sample survey catches once they are on deck are the same as those employed during bottom trawl surveys (Chapter 2). Length data and subsamples to be used for age determination can be collected in the same manner, and data recording and editing procedures will also be identical. When combining length frequency data from several hauls, however, it may not be appropriate to weight each sample on the basis of the catch per unit effort obtained during the haul. Catch rate is often related more to factors such as the success at aiming the trawl where the most dense aggregations were observed, or the use of tactics that minimize gear avoidance rather than being proportional to abundance. In this case, the individual samples should be weighted by the echo counts or echo integrator values obtained in the area immediately adjacent to the depth and area trawled. For any given stratum, the average fraction (f_i) of fish in sex–length group i would then be given by

$$f_i = \frac{\sum_{j=1}^{n} a_j f_{ij}}{\sum_{j=1}^{n} a_j}$$

where f_{ij} = fraction of fish in sample j that are in sex–length group i
a_j = acoustic weighting factor for sample j
n = number of hauls in stratum.

ECHO COUNTING

As mentioned previously, the process of counting echoes from individual fish is limited to situations where these fish are widely separated in space. It is clear that their separation in range from the transducer must be greater than $c\tau/2$ (Fig. 3-6), but there must also be adequate angular separation between the fish. A fish L meters long at range R meters from the transducer will subtend a maximum angle of arctan L/R degrees. This means that a 30-cm fish at a range of 30 m from the transducer would require a beam angle of 0.6° in order to separate its echo from those of any adjacent fish located at the same range. In fact, even the narrowest beams commonly used in echo counting are an order of magnitude wider (6°), and the applicability of echo counting methods consequently depends on the spatial separation of the target species more than the equipment being used. Nevertheless, short pulse lengths, narrow beam angles, and slow boat speeds all improve the probability of detecting single echoes.

It is sometimes difficult to be certain that a given echo return comes from a single fish or is the resultant of multiple echoes from the same sampling volume. Examination of the pulse lengths from returning echoes for returns with durations that deviate significantly from $c\tau/2$ is a useful technique in determining whether a given echo comes from a single fish. Since spatial resolution may vary between different transmissions, it is also useful to follow the echo returns from a suspected single target as the vessel passes above to be certain that they do not resolve into more than one target. Both of these methods can be employed using an oscilloscope.

As discussed earlier, the effective beam width being sampled is a function of the target strength of the fish being sampled, transducer design (source level and directivity), receiver gain, and the threshold level used for counting. Threshold level in turn depends on noise level and the nature of the side lobes produced by the transducer. Noise is present at all depths, and is amplified along with returns from the target population. A 40 log R TVG is used during echo counting, so that the voltage recorded for a given target does not vary with its depth in the water column. However, unamplified echo returns from the target species (signal) do weaken with depth, while background noise remains roughly constant. The TVG amplifies both signal and noise simultaneously, so that a depth will eventually be reached where the signal will no longer rise above the amplified noise level. This depth will depend in part on the quality and design of the echosounder system, and must be sufficiently great to encompass the entire depth range of the target species. If the threshold level is not clearly below the echo level obtained from the smallest fish present, when they are located at the edge of the acoustic beam and in the deepest part of their range, bias will result.

Estimating Volume Sampled

It is possible to determine the effective beam angle knowing the target strength of the species being surveyed, and given careful calibration of the echosounder system (Keiser and Ehrenberg 1990). However, a much simpler and less costly approach is possible using the "duration in beam" technique (Nunnallee 1980, Thorne 1988). This is based on the principle that as the survey vessel passes over a fish, the path traced across the acoustic beam will have an average length determined by the effective sampling angle.

If $\bar{l}$ equals average path length in meters, then $\bar{l}d$ is the area of a rectangle with an area equivalent to the crossection of a circular beam with diameter d (meters), then

$$\bar{l}d = \pi r^2 = \pi\, d^2/4$$

or

$$d = 4\bar{l}/\pi$$

It is also true that

$$\bar{l} = st$$

where s = vessel speed (m/sec)
t = mean duration in beam (sec)

and

$$t = \frac{\bar{e}}{p},$$

where $\bar{e}$ = the mean number of successive echoes obtained from the fish as the vessel passes
p = pulse repetition rate (transmissions or "pings" per sec)

It follows that

$$\bar{l} = \frac{s\bar{e}}{p}, \text{ and } d = \frac{4s\bar{e}}{p\pi} \tag{3.3}$$

Implementation of this technique is relatively simple, and involves collecting the acoustic data on magnetic tape (audio or video) in the field and carefully monitoring vessel speed. Once ashore, the data are processed using an oscilloscope. A threshold is established and data are collected on the number of fish whose echo voltages exceed this threshold, and the mean number of echoes obtained per fish ($\bar{e}$) as they traverse the beam. It is necessary to collect data on echoes per fish from only a subsample (usually about 100) of the fish that are detected. This provides the data required to estimate sampling diameter at a given depth (range) using Equation 3.3.

The data on sampling diameter and range can then be used to estimate the effective beam angle, since

$$\sin(\theta) = \frac{d}{2R} \qquad \text{(Fig. 3-8)}$$

Data from a single depth can be used to estimate θ, but are usually pooled when the investigator is certain that target strength does not vary with depth.

In this case, the $\bar{e}_i$ and R_i data for each fish are fitted to a linear regression model which passes through the origin:

$$\bar{e}_i = bR_i + \varepsilon_i,$$

where the slope $(b) = p\pi \sin(\theta)/(2s)$ and ε_i = random error.

Crittenden et al. (1988) show that an unbiased, minimum variance estimator of $\sin(\theta)$ can be obtained by employing weighted regression, where the weights are $(1/R_i)$. The estimator of $\sin(\theta)$ is then

$$\sin(\theta) = \frac{b2s}{(\pi p)}$$

Sampling volume encompassed by a single transmission between ranges R_1 and R_2 is then estimated by using the equation for the volume of a conic frustum:

$$V = \frac{\pi}{3}(R_2^3 - R_1^3)(\sin\theta)^2\cos(\theta)$$

Estimating Density and Population Size

The estimated volumetric density of fish (D_v, number/m^3) in the volume between R_1 and R_2 is usually obtained simply by dividing the number of observed echoes by the sampling volume, although Crittenden et al. (1988) use bootstrapping to develop an alternative estimator and its variance.

Estimates of D_v along any convenient unit of transect (e.g., 1 km,1 nm, or entire transect) are usually made for several depth strata (Fig. 3-17) and then converted to a single areal (D, number/m^2) estimate of density using

$$D = \sum_{i=1}^{h} D_{vi}(R_{iu} - R_{il}) \tag{3.4}$$

where D_{vi} = number of fish per m^3 in depth stratum i
R_{iu} = upper range limit for depth stratum i
R_{il} = lower range limit for depth stratum i
h = number of depth strata

The most straightforward population estimates for a given geographic stratum consist of treating the mean D values for each transect as T independent observations, suitably weighted to adjust for differences in transect length, and then using the approach outlined in Chapter 1 (Equation 1.1) to obtain population estimates. In this case,

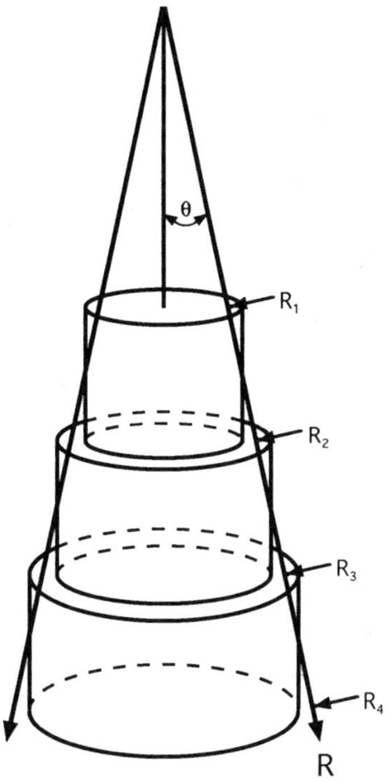

Figure 3-17. Density observations (number/m^3) within the volumes sampled at different depths are summed over the entire water column to determine density per unit surface area (number/m^2).

$$\overline{D}_i = \frac{\sum_{j=1}^{T_i} L_j \overline{D}_{ij}}{\sum_{j=1}^{T_i} L_j} \tag{3.5}$$

where $\overline{D}_i$ = mean density (number/m^2) in stratum i
$\overline{D}_{ij}$ = mean density for the jth transect in stratum i
L_j = length (m) of transect j
T_i = number of transects surveyed in stratum i

The variance for $\overline{D}_i$ can then be derived from the formula for the variance of a ratio estimator for subsampling (Armstrong et al. 1988):

$$\text{Var } \overline{D}_i = \frac{T_i}{T_i - 1} \sum_{j=1}^{T_i} L_j^2 (\overline{D}_{ij} - \overline{D}_i)^2 \bigg/ \left(\sum_{j=1}^{T_i} L_j \right)^2 \tag{3.6}$$

Because of the almost continuous nature of acoustic sampling there are many alternative statistical approaches in survey design, and estimating population size from the data set of observed densities and their spatial locations.

These will be discussed in greater detail in a separate section (see Survey Design and Analysis of Acoustic Data).

However the data are analyzed, the duration in beam method has been shown to be relatively simple to implement and robust with respect to violations in the underlying assumptions. Bias can occur in the upper water column, where the sampling beam is narrow and some fish can pass through the beam without being detected. A simulation study by Nunnallee (1980) showed that this bias becomes significant at values of $\bar{e}$ below 3 successive echoes per fish. Bias can be minimized by employing slow vessel speeds and rapid pulse repetition rates.

Further bias can be introduced if the fish are swimming rather than remaining motionless as the survey vessel passes. Crittenden and Thomas (1987) have extended the duration in beam method to correct for this.

One assumption of the duration in beam method is that the fish being sampled all have the same target strength. However, Thorne (1988) pointed out that the technique is relatively insensitive to variability in target strength due to the nature of the directivity pattern. The average detection angle occurs near the steepest part of the directivity pattern, where the angle changes relatively slowly with target strength changes. In the case of the transducer used in his surveys, a 4-fold variation in the target strength would cause only a 10% change in the detection angle. Thorne found that the sampling angles determined in surveys of juvenile sockeye salmon in 50 surveys had a mean of 18.9°, with a standard deviation of 2.02. All surveys were made using Simrad EY-M echosounders, but were made in 14 different lakes under a wide variety of weather conditions.

Scaling Echo Integrator Outputs Using Counts

When the target species being surveyed is sufficiently dispersed for echo counting in only certain parts of the survey area, it is possible to employ echo integration in those areas where the fish are dense, then use echo counting data from the remaining areas to scale the integrator output. A simple linear regression analysis of the integrator output vs fish density obtained through counts is typically used to analyze the data (Nunnallee 1983). This eliminates the need for the precise calibration and target strength estimation required for echo integration (see next section).

ECHO INTEGRATION

In most applications of acoustics, particularly in marine environments, the fish are not sufficiently separated spatially to allow echo counting. Several individual fish are present in the volume of water being sampled at any given instant (Fig. 3-7) and contribute to the echo signal returning to the survey vessel. As a result, it is necessary to use Equation 3.2 and sum (or integrate) the echo intensities or squared voltages (V^2) over the depth range being sampled. At any given range (R), the intensity being received at the transducer from a density of D_v fish per unit volume is given by

$$I = \frac{D_v\,(c\tau/2)\psi\; kI_0 e^{-2\beta R}}{R^2}$$

Echo integration systems apply a TVG correction, and integrate the V^2 values over a given depth range ($R_2 - R_1$), so that the output (M) for a given depth range is

$$M = \int_{R_1}^{R_2} R^2 e^{2\beta R}\; Irga = \int_{R_1}^{R_2} V^2 R^2 e^{2\beta R}$$

where r = receiving sensitivity (or voltage response) of the transducer
g = total gain of the system
a = TVG correction factor (to correct for deviations from ideal TVG)

Given accurate calibration of the system, M can then be related to the number of fish per unit volume (D_v) in depth interval ($R_2 - R_1$) by

$$D_v = M/K$$

where K = the constant required for the integration process:

$$K = (c\tau/2)\psi\; kI_0 rga$$

The D_v values for each depth interval ($R_2 - R_1$) can then be summed over the entire water column using the same procedure used in echo counting (Equation 3.4), to estimate the number of fish/m² (D).

Since acoustic energy from all sources, including reverberation (echo returns) from nontarget species, gas bubbles, suspended particles, and plankton is being integrated over the interval $R_2 - R_1$, it is critical that these integration intervals be selected with care. The investigator must examine the echograms and net sampling data carefully and eliminate any depth strata where significant concentrations of nontarget species or plankton are present, so as to restrict the D_v values to those originating from the target species. Reverberation from the bottom produces a much greater level of acoustic energy than that from concentrations of fish, and can be a major problem when the target species is near the bottom. Bottom reverberation can cause serious errors if not screened out, and "bottom tracking" features are available on some integration systems to aid in the screening process. However, bottom tracking is complex when the bottom is rough, where the slope is steep, or where the transducer is rolling because of bad weather, and it is often necessary to override the automatic system.

While it is a common practice to define the integration intervals manually, the Bergen Echo Integrator system (Knudsen 1990) includes a feature whereby a mouse-operated cursor can be used to sketch out integration intervals on the echogram display (Fig. 3-11). Different layers can be defined using the cursor, and assigned different species compositions. The cursor can also be used to define the bottom in those cases where the operator desires to override the bottom tracking algorithm.

No matter how carefully the integration intervals are defined, however, any nontarget reverberation present in the depth interval selected will be included in the D_v values. If a plankton layer extends into the integration interval, an integration value can be obtained from an area outside the zone where fish are present, and subtracted from the D_v value for fish. Dense plankton recordings may occasionally mask the fish recordings, however, and make it impossible to use the observations quantitatively.

Noise presents further problems, since it is present in all D_v values unless adjusted for. A common practice during echo integration is to simply apply a threshold level to the voltages before they are integrated. Because of the TVG, the threshold level should be proportional to $R^2e^{2\beta R}$ in order to remove a constant noise level. This threshold sets a limit to how far out in the beam a single target is allowed to contribute to the integrator value, and must be low enough to allow the smallest targets to contribute throughout the most important part of the beam. The threshold effect is determined by noise, and is negligible only when the smallest individuals of interest give an echo that exceeds the integrator threshold level in all positions within the significant part of the beam. Ideally, this should be the case even when the fish is at its maximum depth and presenting its least favorable aspect to the transducer. If this is not the case, serious error can result. When the target species is dispersed and sparsely distributed, and a low performance system is used, over 90% of the echo energy can be lost due to the threshold effect (Aglund 1989). For this reason acoustic engineers and scientists have placed a premium on the development of low-noise echo integrators (Knudsen 1990), transducer mountings, and survey vessels.

Often, no corrections are made for noise when it overlaps fish echoes that rise above the threshold. Nunnallee (1990) recommends that the noise level be measured when the transducer is not transmitting (in the receiving mode only) then subtracted from the integration totals.

Acoustic energy can be lost due to attenuation by either air bubbles or, when the target species is found in dense schools, by fish themselves. This loss of energy always results in underestimation of abundance, and must either be avoided or corrected for. With most conventional hull-mounted transducers, air bubbles can be entrained along the hull and flow over the transducer face as a result of the normal movement of the survey vessel through the water. This source of error can be avoided with careful location and mounting of the transducer.

Air bubbles in the surface layer that are the result of wind and waves are more difficult to avoid, however, and their presence can cause serious errors. One way to reduce this error is to mount the transducer in a hydrodynamically stable body, and tow it below the layer where air bubble concentrations are significant. In many cases, the survey is simply stopped during stormy weather. This is a good practice, not only because of the problem with air bubbles, but because rolling and pitching of the transducer also increases in stormy seas. When the orientation of the transducer varies significantly between transmission and reception of a pulse, the directivity pattern can depart significantly from what is predicted.

Attenuation of acoustic energy within a school (shading) can result in a failure for some of the energy transmitted to reach fish located in the deeper parts of the school. Attenuation by schools depends on the density and vertical extent of the schools being observed, and is sometimes indicated by a significant reduction (or even loss) of the bottom echo. This appears to be a problem only when schools have a vertical extent greater than about 20 m at the densities (1–2 fish or 0.5 kg/m^3) commonly observed in fish schools (Simmonds et al. 1992). Since most fish schools tend to disperse at night, the problem can often be avoided by surveying then. If the problem still remains, it may be possible to correct for it if the extinction cross section is known for the target species (Foote 1990).

It should be clear at this point that careful design and calibration of the echo integration system, careful survey design, and constant attention to detail are imperative if an echo integration survey is to be successful. When done properly, however, the sampling intensity provided by acoustics allows an intensity of coverage and level of spatial detail that is unrivaled by any other direct survey method.

A separate observation of D (number/m^2) is available from each transmission, and can be used in a variety of ways to obtain population estimates for the survey area. As with echo counting, the most straightforward of these is to treat the mean density observation for each transect as an independent estimate, and use Equations 3.5 and 3.6 to obtain population and variance estimates for the survey area. There are a number of alternative ways to make these estimates, however, and these will be covered in a separate section on the statistical treatment of acoustic data.

Calibration

Recall that the density observations (D_v, number/m^3) for any depth interval are related to the integrator output (M) by

$$D_v = M/K$$

where $K = (c\tau/2)\psi k I_0 r g a$.

The k, or target strength parameter, should usually be determined during the survey itself and will be discussed in the next section. The goal of calibration, then, is to determine the remaining parameters in K precisely. Calibration experiments are usually carried out before the survey, and then repeated at convenient intervals during the course of the survey to be certain that system parameters are stable.

While a number of calibration methods, including measurements of known quantities of caged fish, reciprocity calibration using two transducers, and the use of calibrated hydrophones have been used in the past (MacLennan and Simmonds 1992), it is now generally accepted that the use of standard calibration targets provides the most accurate measurements of system parameters. Standard copper or tungsten carbide calibration spheres with well-defined acoustic reflection properties have become available, and techniques for using them have been well-standardized (Foote et al. 1987) for use with split beam and dual beam transducers (Fig. 3-23).

By placing a calibration sphere in a harness that allows careful control of its position in the acoustic beam (Fig. 3-18), it is possible to place it directly on the acoustic axis at a known range. Integration of echo returns from the sphere can then be used to determine the product of source level (I_0) and on-axis receiving sensitivity (r). This parameter reflects the combined performance of the transmitter, transducer, receiver (at a given gain setting), and integrator simultaneously. Thus the amplitude, center frequency and

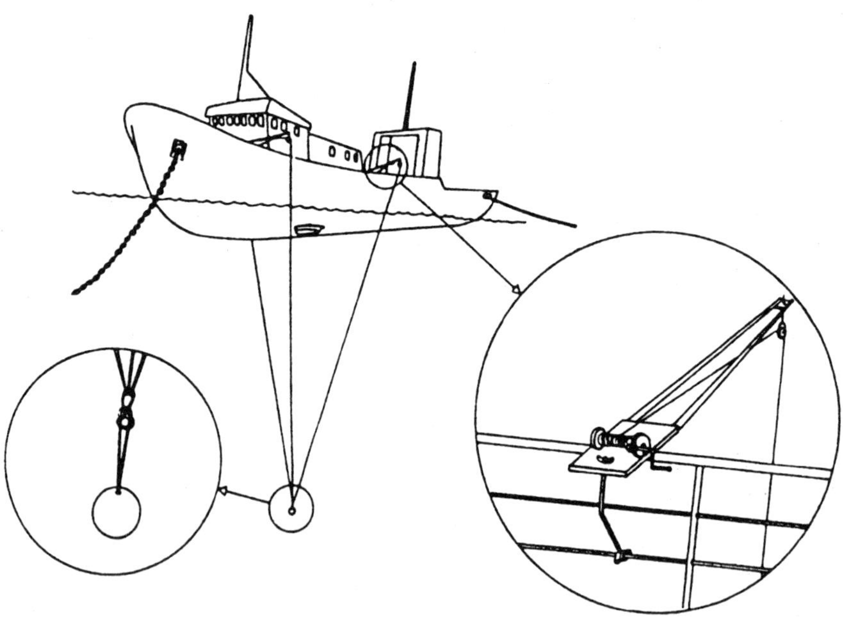

Figure 3-18. Rigging of an acoustic survey vessel for calibration with a stationary sphere. From Foote et al. (1987).

duration of the transmit signal, transducer bandwidth and sensitivity, receiver bandwidth and gain, and performance of the echo integrator are all taken into account. Calm water is obviously required to carry this process out, and it is a common practice for the survey vessel to make several preplanned stops in sheltered bays for purposes of calibration.

By moving the calibration sphere systematically along arcs in several planes through the acoustic axis, the beam pattern for a split beam transducer can be measured (Fig. 3-19), and used to determine ψ. This procedure requires considerable effort, however, and since ψ usually varies little over time, it is carried out only infrequently. The beam pattern values obtained by the man-

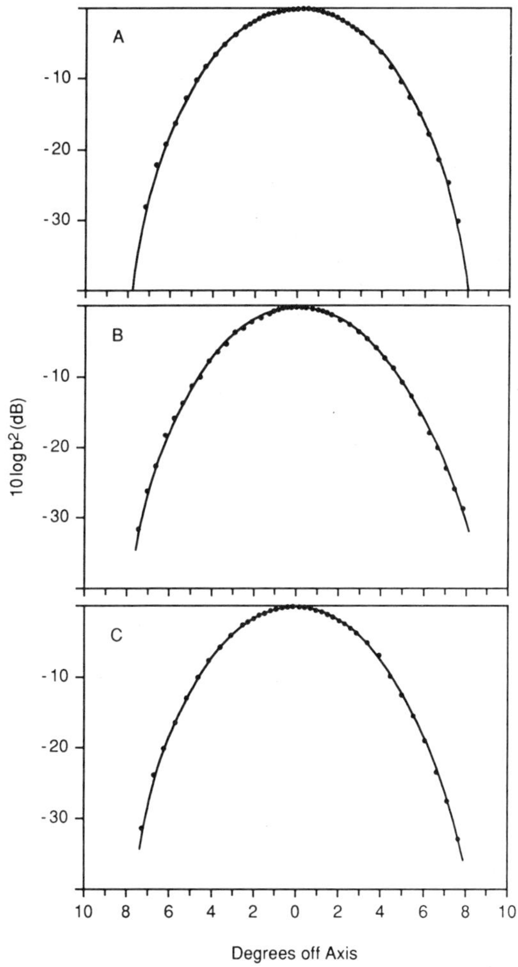

Figure 3-19. Two-way directivity patterns [10 log $b^2(\theta)$ versus degrees off acoustic axis] through three cross-sections for a 38-kHz split beam transducer, determined using a standard sphere. (A) 62–222°, (B) 146–326°, (C) 90–270°. From Foote et al. (1987).

ufacturer can differ substantially from those obtained after the transducer has been mounted on the hull or in a towed body (Foote et al. 1987), and ψ should be determined at least once after mounting.

The TVG function typically deviates measurably and not insignificantly from the ideal specification. Correction of the TVG function (a) often requires instrumentation external to the echo-sounding and echo-integrating equipment, with signals being applied to the amplifier input, and the resulting output being measured. Some echo integrators are equipped with special signal generators that produce a known signal at a simulated depth and can be used to monitor TVG performance at frequent intervals. This same approach can be used to examine the relationship between input signals and output from the integrator to be certain that it is linear and not distorted by limitations in the dynamic range of the system.

Most modern integrators are designed so that changes in the speed of sound (c) can be input as the user sees fit. This is particularly important during calibration, and should be calculated from temperature–salinity profiles collected at the calibration site. MacLennan and Simmonds (1992) provide equations for estimating c from temperature, salinity, and depth.

The effective pulse length (τ_e) is determined by a number of factors, including the nature of the bandpass filters used in signal processing. Significant error can be introduced through use of the manufacturer's nominal value for τ_e, and Traynor et al. (1990) recommend that this be measured from individual fish targets, where all factors affecting pulse length are included.

Calibration results should be carefully recorded, and reported with any survey results (Table 3-1), along with information on gains and thresholds used during the survey. While calibration requires strict attention to detail, procedures have been refined to the point where overall error due to calibration can be kept to within $\pm 3\%$ (Foote et al. 1987). This makes it one of the most accurate components of the echo integration process.

Decibels and the Sonar Equation

The integration equation and calibration parameters are frequently expressed in terms of decibels (dB), which simply measure relative intensity, i.e.,

$$\text{dB} = 10 \log \left(\frac{I_1}{I_2}\right)$$

where I is intensity (watts/m^2).

For example, a halving of echo intensity would correspond to a 10 log (0.5) or -3 dB change, while a doubling would mean a 10 log (2) or 3 dB increase.

Often I_2 is a standard level of intensity, usually referenced to a pressure of 1 micropascal (μPa). When this is the case, the appropriate notation for this should be, for example, source level (SL) = 200 dB/μPa measured at 1 m, and receiving sensitivity (or voltage response, VR) = -100 dB/V/μPa. A corresponding notation seen in the literature is SL = 200 dB//1 μPa at 1

Table 3-1. Documentation for acoustic systems

Echosounder
Manufacturer and model
Transmitter power (kW)
Operating frequency (kHz)
Bandwidth (kHz)
Effective pulse length (ms)
Transducer mounting: Stabilization, towed body design, etc.
Angles between −3 dB points (degrees)
ψ
Source level, SL (dB/μPa at 1 m)
Voltage response, VR (dB/V/μPa) or SL + VR (dB)
TVG: 20 log R or 40 log R, α applied
Overall gain (dB)
c (m/sec)
Integrator
Manufacturer and model
Overall gain (dB)
Threshold applied (V or dB)
Vessel, trawl design, and deployment
As in Table 2-1

m and VR = −100 dB//1 V per μPa. Often the reference unit is implied rather than being stated specifically, although this can easily be misinterpreted if the reference used is 1 microbar (1 μb = 10^5 μPa).

In decibel notation, the general equation for echo intensity returning from a target at a given range R

$$I = \frac{I_0 k b^2(\theta, \phi) e^{-2\beta R}}{R^4}$$

becomes

$$\mathrm{EL} = \mathrm{SL} + \mathrm{TS} + 20 \log b(\theta, \phi) - 2\alpha R - 40 \log R$$

and is known as the "sonar equation," where SL = source level and TS = target strength. The last two terms are commonly used to describe the two-way time varied gain (TVG) correction for an echo sounder. Notice that attenuation is given in terms of α (dB/m), where $\alpha = 10(\log_{10} e)(\beta) = 4.3\beta$.

Target Strength

As mentioned previously, the swim bladder is often responsible for 90–95% of the energy reflected by those fish that have them. Most fish have swim bladders, although flatfish, elasmobranchs, sand lances (Ammodytes), and

Atlantic mackerel are notable exceptions. Some bathypelagic species have a swim bladder filled with oil or invested with fat, so that their target strength will be reduced considerably from an air-filled bladder of comparable size.

For any given species, the surface area (dorsal aspect) of the swim bladder increases approximately as the square of length, and for species with swim bladders, a rough idea of target strength can be obtained knowing their average length. Target strength also increases with the frequency of the echosounder being used, and equations have been developed for many species that relate target strength to length and wavelength (λ). Love (1971), for example, used data on dorsal-aspect target strength obtained for a variety of species and frequencies, and fit the data to the regression equation:

$$\text{TS} = 19.1 \log L + 0.9 \log \lambda - 23.9 \text{ dB}$$

where L and λ are both measured in meters.

These relationships vary substantially between species, however, and merging of target strength data across species and frequencies is generally unjustified (Foote 1979). Wide differences in target strength can also occur within a species due to the aspect presented to the echosounder (tilt angle), or to changes in swim bladder volume associated with vertical migrations, feeding (stomach distension), or gonad maturation. In herring, there is an inverse relationship between fat content of the body and swim bladder volume (Ona 1990).

An important distinction in understanding the effect of bathymetric migrations on changes in swim bladder volume exists between physostomes and physoclists. Physoclists (e.g., gadoids) have a closed swim bladder with no ducts to the exterior, together with a gas gland for secretion and an occlusable "oval window" for gas resorption. Both mechanisms operate slowly (secretion is particularly slow), and compensation for depth changes will consequently lag behind the rate of vertical migration (Table 3-2). Deviations from neutral buoyancy result, and must be compensated for behaviorally, through changes in tilt angle, swimming speed, and extension of the pectoral fins.

Physostomes (e.g., clupeoids, salmonids, and cyprinids) usually have an open swim bladder with a duct (or ducts) to the exterior, and typically lack a gas secreting mechanism. Most physostomes obtain gas by swallowing air at the surface, and will not be neutrally buoyant at any depth unless they are positively buoyant at the surface. Herring have been observed to release gas bubbles after completing their upward migration at dusk (Thorne and Thomas 1990), and these authors conclude that the swim bladder is probably refilled at dawn, prior to downward migration to some target depth where they are neutrally buoyant. Gas release by herring was also observed in strong tidal currents.

Behavioral changes in aspect are known to cause wide variations in target strength, both in experimental and field situations. A number of experiments have been conducted with caged fish, whereby target strength estimation and

Table 3-2. Reported maximum vertical excursions by fish and time required to adapt to neutral buoyancy

Species	Diel Change in Depth (m)	Pressure Change (%)	Time Required to Adapt at 10°C (hours)
Cod	203→148	−26	1.27
	148→203	+35	55
Cod	190→139	−26	1.25
	139→190	+34	51
Cod	128→108	−14	0.67
	108→128	+17	20
Haddock	170→110	−33	1.7[b]
	110→170	+50	60[b]
Saithe	160→60	−59	3.75[b]
	60→160	+143	100[b]
Perch	20→8[a] 10→2[a]	−40	8

[a]Hypothetical examples.
[b]Assuming gas secretion and absorption same as cod.
Source: Blaxter and Batty (1990).

television monitoring are carried out concurrently, and the results (Fig. 3-20, for example) reveal the potential for wide variation in target strength from variations in tilt angle alone. Diel variation in orientation can also occur, with a tendency for mean tilt angles to be positive and larger at night (Table 3-3, Blaxter and Batty 1990).

Data on tilt angles are difficult to obtain in the field, but repeated target strength observations on single fish (Fig. 3-21, for example), as well as the high variances that typify target strength data in general (Fig. 3-22), indicate that variations in individual orientation are extremely important. Traynor and Williamson (1983) found that the target strength of walleye pollock was about 3 dB lower at night than during the day, and attributed this to variation in mean tilt angle.

Estimating Target Strength

Variations in swim bladder volume and tilt angle have been shown to cause substantial variation in target strength. Research has shown clearly that target strength must be measured at the same time and location (*in situ*) that echo integration data are being collected if bias is to be avoided. As a result, any techniques of target strength estimation that rely on observations of caged, tethered, or dead fish are of only limited use. Similarly, the method of making casts of the swim bladders for several individuals, and using them to model and reconstruct the scattering field for the species (Foote and Ona 1985), can

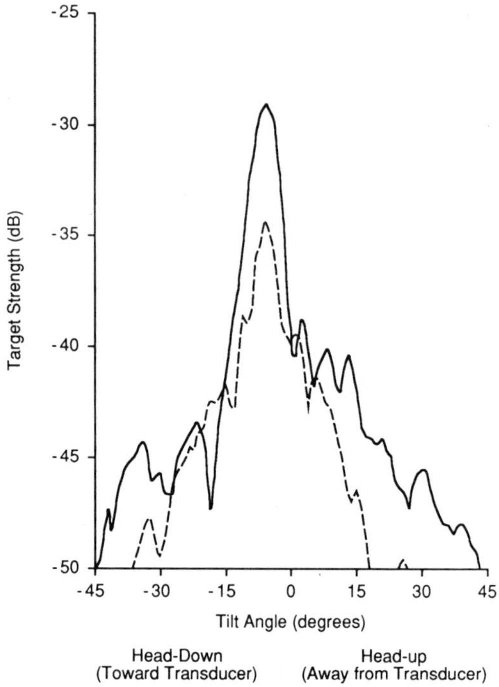

Figure 3-20. Ventral-aspect target strength measurements for encaged saithe (*Pollachius virens*) with a mean length of 35.1 cm (±0.6 cm). The solid line represents the average values for 16 individuals observed at 38 kHz, the dashed line values for 17 observed at 120 kHz. From Foote (1978).

Table 3-3. Tilt angle of fish under different conditions[a]

Species	Depth (m)	Location	Tilt Angle Degrees (±SD) Day	Night
Saithe	4	Large pen	−0.9(±5.4)	—
Capelin	44	Sea	3.3(±18.4)	—
Herring	2–5	Large pen	−3.2(±13.6)	3.8(±6.0)
Herring	20	Sea	−3.4(±10.3)	12.0(±23.5)
Herring	2.5	Small cage	2.9(±14.2)	—
			−3.1(±11.5)	—
Herring	1.5	Small pen	−3.9(±12.8)	—
	4	Large cage	−0.2(±11.9)	—
	30	Large cage	8.1(±16.9)	—
Mackerel	?17.5	Cage	8.8	18.1[b]
		Cage	8.3	12.3[b]

[a]Negative values indicate head below the horizontal.
[b]Tilt angle significantly higher during the night ($P<0.01$).
Source: Blaxter and Batty (1990).

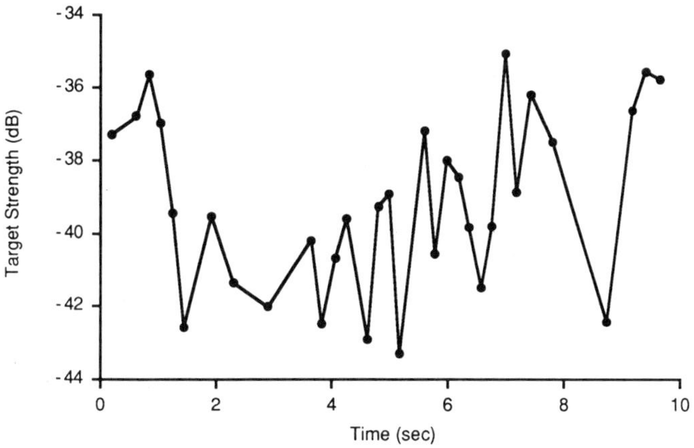

Figure 3-21. Observations, probably for a striped bass (*Morone saxatilus*), of night-time dorsal-aspect target strength. Observed in Lake Mead using a 420-kHz dual beam transducer. From Dawson and Karp (1990).

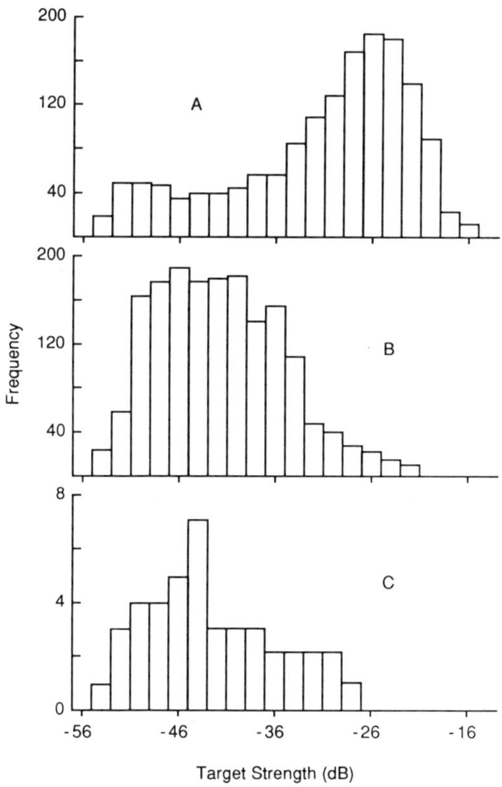

Figure 3-22. Frequency distribution for target strengths of (A) cod, (B) capelin, and (C) mackerel measured at Brador Bay, Canada. From Rose and Leggett (1988).

provide only approximate estimates of *in situ* target strength. Exact estimates would be possible only if *in situ* data on tilt angles and swim bladder volume were available. As a result, *in situ* estimates of target strength have become an integral part of echo integration surveys.

The basic problem in estimating target strength lies in the fact that the echo intensity received from a given fish is dependent on its location in the acoustic beam (θ, ϕ) as well as its target strength. Some means must therefore be devised to separate the effects of an individual's target strength and location in the acoustic beam on its echo returns.

In situ target strength estimates are usually made directly, using special equipment to either eliminate the effect of the beam pattern (dual-beam technique) or to determine the location (θ, ϕ) of a target in the beam (split-beam technique). It is imperative that single-fish echoes are isolated, so that the requirements for direct target strength estimation are the same as those for echo counting. The targets must be sufficiently far apart that multiple echoes do not occur, and when schools are dense it is often impossible to obtain target strength data for the same aggregations of fish being echo integrated. However, the use of short pulse lengths, narrow beam angles, and slow vessel speeds improves the chances of detecting single echoes. Many of the analytical techniques used to define single targets in echo counting (criteria for minimum acceptable pulse duration, target tracking over multiple transmissions, etc.) are also used to identify single-fish targets for the estimation of target strength. Target strength estimates are usually taken for several thousand individual fish, and their mean and variance obtained. It is important that the signal-to-noise ratios be sufficiently high that the smaller individuals are not eliminated from the sample because of threshold limitations.

The dual-beam technique employs a transducer constructed such that the entire array transmits in a narrow beam. In reception, the entire array again forms the narrow beam, while a small central array of elements simultaneously forms a wide beam. The expected voltage received from a given target on the narrow beam channel is given by

$$V_n^2 = K_1 b_n^2(\theta, \phi)k$$

while the voltage on the wide beam channel is given by

$$V_w^2 = K_2 b_n(\theta, \phi) b_w(\theta, \phi)k$$

where K_1 and K_2 = calibration constants
k = the target strength

Originally, systems were engineered so that $b_w(\theta, \phi) = 1.0$ over the entire range of the narrow beam (Fig. 3-23), and the ratio of $V_n^2/V_w^2 = K_1 b_n(\theta, \phi)/K_2$. This could then be solved for $b_n(\theta, \phi)$, and the result substituted into the original equation for V_n^2 to obtain target strength (k). However, dual beam

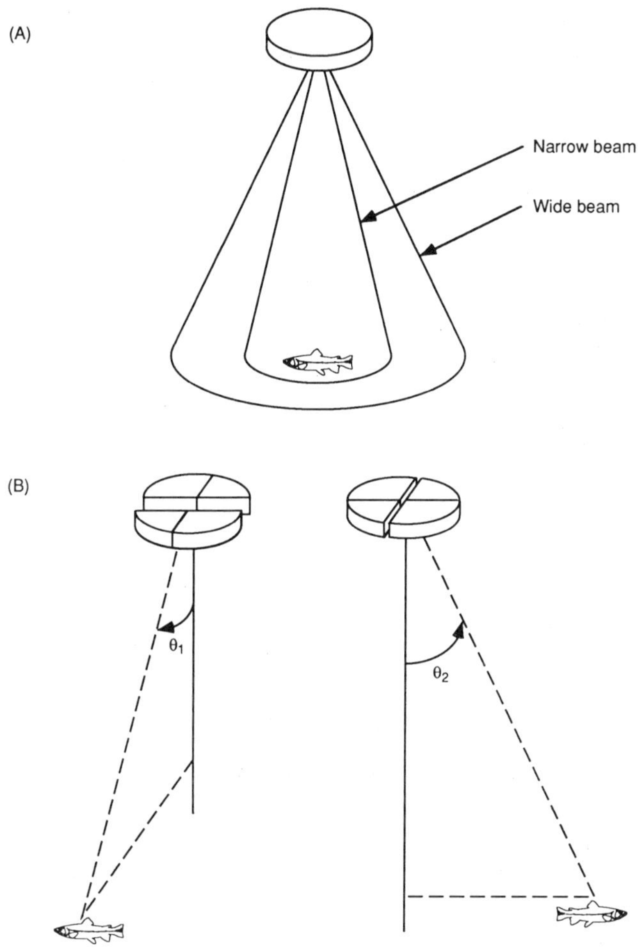

Figure 3-23. Direct target strength estimation using dual beam (A) and split beam (B) transducers.

theory has also been extended to systems where $b_w(\theta, \phi)$ is not unity over the entire range of the narrow beam (Kautsky 1989).

The split-beam technique is based on the division of the transducer into four quadrants. All act in concert during transmission, but each quadrant forms its own beam during reception. Summing the quadrant beams to form fore-aft and port-starboard half-beams and comparing the phase differences between them allow the angular position (θ, ϕ) of the target to be obtained (Fig. 3-23). Since $b^2(\theta, \phi)$ for each target can be determined from the calibration data for the directivity pattern, target strength (k) can be estimated directly from the V^2 data. The split-beam technique is more difficult to implement than the dual beam, due to the nature of the hardware and/or software required for measuring phase differences. Nevertheless, it performs better than the dual-beam system in the presence of noise (Ehrenberg 1983, Traynor and Ehrenberg 1990).

A wide variety of indirect *in situ* methods, which require only a standard single-beam transducer, have also been developed (Ehrenberg 1983; Foote 1991). These methods are based on the use of different statistical procedures to infer the target strength distribution given calibration data for $b^2(\theta, \phi)$ and a data set of V^2 observations for single-fish echoes. Although some successes have been reported, indirect techniques are often susceptible to numerical and statistical error and do not work well in many cases of interest (Ehrenberg 1983).

The vessel avoidance responses discussed in Chapter 1 have a special significance during *in situ* estimation of target strength, in that sudden changes in tilt angle may be associated with the avoidance response. When fish detect an approaching vessel, they typically turn away from the vessel, gradually increase their swimming speed, and begin to dive (Olsen 1990). These responses can result in substantial reductions in target strength when they occur, the exact extent depending primarily on the change in tilt angle that accompanies the diving reaction. As a result, it is critical that the echogram be examined carefully for evidence of diving responses during both integration and target strength estimation. Special studies, using a second vessel and/or long-range multibeam sonar (Chapter 5), are probably necessary to be certain that avoidance responses that occur well in advance of the vessel are not compromising survey results.

It is always critical that the behavior and tilt-angle distributions of the target species are similar during both the echo integration and target strength estimation segments of a survey. This is often difficult owing to the need to work with sparsely distributed individuals during target strength estimation and to isolate single targets. In the case of densely schooling species such as Pacific hake off the coast of Washington, Oregon and California, and walleye pollock in the Bering Sea and Gulf of Alaska, for example, assessment scientists must often carry out integration surveys during the day and target strength estimation at night when the schools disperse. Experience has shown that special day–night target strength studies are required to explore the bias that can result from differences in behavior pattern during these periods and to make suitable corrections in the data (Traynor and Williamson 1983).

Because of the difficulties involved in obtaining *in situ* target strength data during each survey, many practitioners use *in situ* data to develop a length–TS relationship based on accumulated data. This is then converted to a length–acoustic cross section (σ) relationship, and used in conjunction with length–frequency data from trawl samples to estimate the average acoustic cross section ($\overline{\sigma}$) for the area in question (MacLennan and Simmonds 1992). Mean target strength (TS) can then be calculated from the relation TS $= 10 \log(\overline{\sigma}/4\pi)$.

As a final note, while target strength is always estimated for individual fish, it is often expressed in terms of TS/g when it is desired to estimate the biomass (e.g., g, kg, mt) rather than numbers of fish in the survey area. Conversion between a numerical (TS_n) and weight basis (TS_w) for target strength is possible once the TS–length and length–weight relationships are known.

The length-weight relationship usually takes the form

$$W = aL^b$$

where W = weight (g)
L = length (cm)
a and b = constants

Individual TS (TS_n) is usually related to length by

$$TS_n = c + d \log L$$

and TS/g (TS_w) by

$$TS_w = e + f \log L$$

Since the weight of an individual fish of length L (in decibels) is 10 log (aL^b), then

$$TS_w = \frac{TS_n}{10 \log(aL^b)} = e + f \log L$$

or

$$c + d \log L - 10 \log a - 10\, b \log L = e + f \log L$$

From this it can be seen that the constants can be related by

$$e = c - 10 \log a$$

and

$$f = d - 10\, b$$

SURVEY DESIGN AND ANALYSIS OF ACOUSTIC DATA

Survey Design

Because of the almost continuous nature of acoustic sampling, a number of unique problems and opportunities exist in terms of survey design. A recent synopsis of the practices used in acoustic survey design and analysis (Simmonds et al. 1992) shows that a wide array of techniques are employed, and that opinions often differ as to which approach is best.

One example lies in the issue of whether to use discrete transects as the basic sampling unit, or to continue collecting data at all times during the survey, resulting in a continuous trackline (Fig. 3-24). While continuous tracklines have the benefit of using all possible data that can be collected, however, they bear a hidden cost due to the fact that the survey area is not covered in a uniform manner. In the case of zig-zag designs, for example, there is a higher sampling intensity at the turns compared to other segments of the trackline. This exacerbates a problem characteristic of all acoustic surveys, namely nonindependence and autocorrelation between density observations from adjacent portions of the trackline.

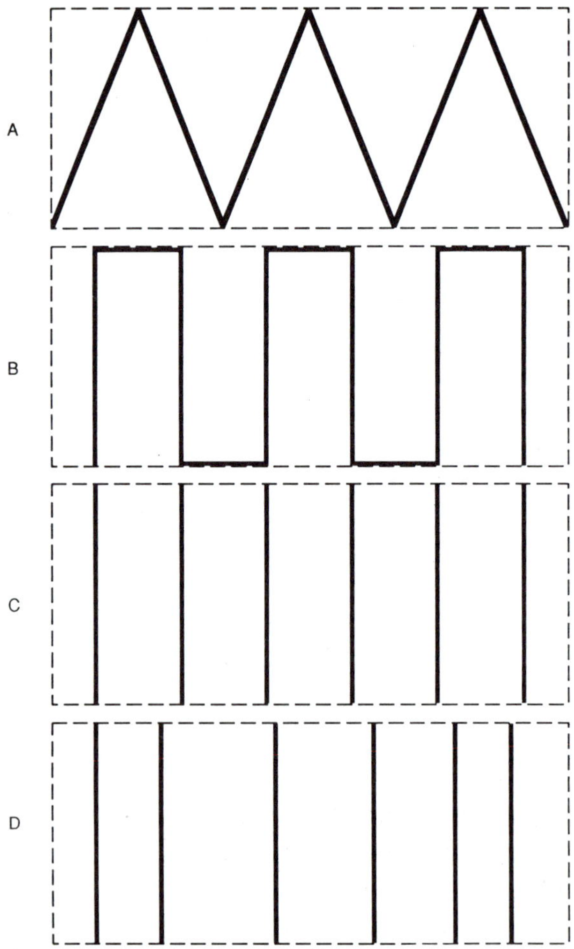

Figure 3-24. Trackline patterns used during acoustic surveys. (A) Continuous zigzag, (B) continuous parallel, (C) discrete systematic parallel, (D) discrete random parallel.

This characteristic makes it impossible to compute unbiased variance estimates without carefully evaluating autocorrelation as a function of distance between transect segments (MacLennan and MacKenzie 1988), and determining a spacing beyond which autocorrelation does not exist. Alternatively, any desired length of transect distance (e.g., 1 nm) or time (e.g., 1 minute) can be used as the sample element, and the variance computed by treating the results as a cluster sample (Shotton and Bazigos 1984). Williamson (1982) treated the individual density observations on each transect segment as a cluster of homogeneous observations and allowed for within- and between-cluster components of the variance. This method has been shown to yield reliable confidence intervals with certain simulated fish distributions. However, Francis (1985) pointed out that the cluster sample approach depends on stationarity (i.e., individual density observations from different parts of the survey area have the same expected value), a situation that is seldom the case in actual surveys. When the distribution is nonstationary, the cluster sample method underestimates the true variance.

Francis (1985) suggested that survey designs be based on sample elements consisting of discrete transects to minimize autocorrelation. Jolly and Hampton (1990) also advocated the use of discrete transects as sample elements, and suggested they be placed quasirandomly within a given stratum. They employed a two-stage procedure, whereby each stratum was divided into a number of strips of equal width running parallel to the direction of the transects. These strips were designated as first-stage units and transects as second-stage units. When strips are selected at random within the stratum, and one transect selected at random within each strip, estimation of the mean density and its variance for each stratum follows the same procedure outlined in the section on echo counting (Equations 3.5 and 3.6) in this chapter, while estimation of population size and its variance are the same as in Chapter 1 (Equations 1.1 and 1.2).

There still remains a chance, however, of two sample transects occurring close to each other, and Jolly and Hampton discarded any transect lying within one strip width of a previously selected transect. It was concluded that this rule would have a negligible effect on theoretical considerations. Figure 3-25 shows the initial transect plan developed for an acoustic survey of anchovies off South Africa by Jolly and Hampton. Allocation of sampling effort between strata was based on optimum allocation, using variance estimates obtained from a previous survey.

Francis (1985) suggested that transects be uniformly spaced, since they will then (on average) be separated by the greatest possible distance, minimizing the chance of intertransect correlation. Many practitioners consider the simplified logistics associated with equally spaced transects to outweigh concerns over nonrandom placement (Simmonds et al. 1992).

Repeated systematic surveys, where the entire survey design is replicated several times, have been used to examine the impact of autocorrelation (and other violations of assumptions) when using equations such as Equation 3.6

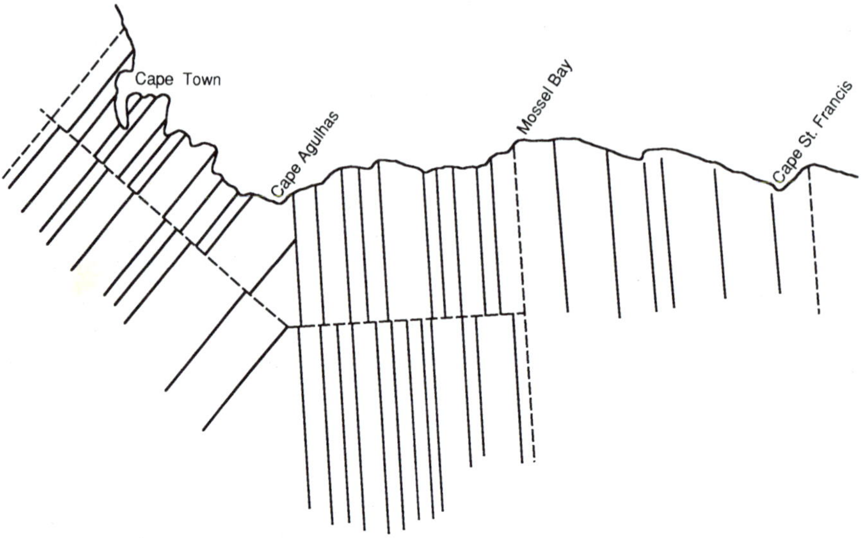

Figure 3-25. Initial transect plan for an acoustic survey of anchovies off South Africa. Dashed lines indicate stratum boundaries. From Jolly and Hampton (1990).

to estimate variances and confidence intervals (Aglen 1989). Aglen assumed that the average density estimates for each transect were independently distributed random variables, and used equations similar to 3.5 and 3.6 (taking all $L_j = 1.0$) to estimate the coefficient of variation ($\mathrm{CV} = \sqrt{\mathrm{Var}\,\overline{D}}/\overline{D}$) for each of the surveys in Table 3-4. These surveys were repeated several times, and equations similar to 1.3 and 1.4 were used to estimate the variance for repeated surveys.

Table 3-4. Repeated systematic surveys reviewed by Aglen (1989)

Area[a]	Predominant Fish Species	Size of Area Covered (sq. n. miles)
Fjellangervagen	Sprat	0.17
Straumsosen + Spjeldnesosen	Herring	1.0
Outer Eidfjord	Sprat	13
Samlafjord	Sprat	16
Nordfjord	Sprat	22
Eidfjord	Sprat	25
Lofoten	Cod	120
Gulf of Oman	Lanternfish	10,000
Barents Sea	Capelin	50,000

[a]Fjellangervagen, Straumsosen, Spjeldnesosen, Eidfjord, Samlafjord, and Nordfjord are fjord areas on the west coast of Norway.

Analysis of autocorrelation between transects suggested some positive dependence between neighboring transects at spacings ranging from 0.06 to 25 nautical miles. Although any such autocorrelation between sample elements (transects) should have resulted in underestimates of the variance when using Equation 3.6, a comparison with CV estimates obtained using the equations for repeated systematic sampling (Fig. 3-26) gave no evidence of bias.

Two-phase designs (Chapter 1), whereby part (e.g., 25%) of the survey effort is held in reserve during the first part of the survey (phase 1), have also been employed in acoustic surveys. The reserve effort is then allocated between strata during a second sweep (phase 2) of the survey area on the basis of the results from phase 1. Exact procedures for allocating effort differ (Jolly and Hampton 1990, Francis 1991). This approach could offer some potential for reducing the variance for population estimates provided that the distribution of the target species is relatively stable between phase 1 and phase 2 of the survey. Care needs to be taken to avoid extensive running time between transects during the second phase of the survey, however (Jolly and Hampton 1990).

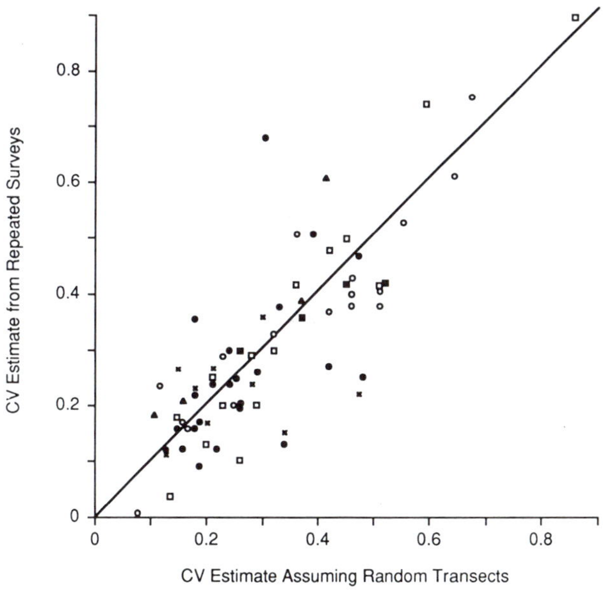

Figure 3-26. Comparison of coefficients of variation (CV) obtained by using the variance estimator for repeated systematic surveys and by treating each transect as an independent random variable. Exact agreement is denoted by the solid line; points indicate individual observations from the survey areas in Table 3-4. From Aglen (1989).

Data Analysis

Because of the extensive coverage of the survey area afforded by acoustic sampling, many researchers prefer to contour the data into areas of isodensity as a first step in data analysis. As a general rule, 3–5 density levels are defined, poststratifying on the basis of the survey data. The results depend on the intuition and analytical procedures used by the person doing the contouring, and can be quite subjective (Simmonds et al. 1992).

While this approach is useful in illustrating distribution patterns for graphic purposes, serious problems arise when the density groups are treated as strata, and variance calculated using standard formulas (e.g., Equation 1.2). Stratification on the basis of the variable being estimated (density) alone clearly results in biased variance estimates. This can be illustrated by the extreme case, whereby the data are subdivided into the maximum possible number of strata with only two observations per stratum. Through careful selection, the density observations within each stratum could be made so similar that the variance estimate for the survey area would approach zero.

Attempts have also been made to apply some of the methods used in spatial statistics to the analysis of acoustic data (Simmonds et al. 1992). While classical statistics attempts to address spatial structure (e.g., clines and gradients) through stratification, some practitioners of spatial statistics prefer to model the data using a trend surface, employing polynomial regression for example. The model is then used to calculate a mean and variance for the data. In theory, where well-defined gradients (onshore–offshore, latitudinal, etc.) exist, an analysis that maps this structure accurately should also result in a reduction in the variance.

A variety of approaches have also been taken to kriging acoustic data (Simmonds et al. 1992, Sullivan 1991). Theoretically, this is the method of choice where density observations are autocorrelated, although research into the practical application of this technique is still in progress. Kriging can result in reduced variance estimates when the appropriate error structure is taken into account. One problem is that it is necessary that the distributions are stationary (i.e., have a constant expected value at different locations within the survey area) in order to employ most types of kriging and this requirement is seldom met in field situations. However, ways to address this problem do exist, such as stratifying the survey area into areas where trends in abundance do not exist, and then kriging within strata.

EXAMPLES OF SUCCESSFUL APPLICATIONS

Echo Counting

Applications of echo counting techniques are most common in freshwater lakes, where fish densities are often low enough to enable single targets to be readily isolated. Acoustic surveys of Lake Tustamena, Alaska, offer one of many possible examples of the use of echo counting in estimating salmonid abundance in oligotrophic lakes. Lake Tustamena is a glacial lake on the Kenai peninsula, and is 40 km long and 8 km wide with a maximum depth of 300 m. The lake is a nursery area for sockeye salmon (*Oncorhynchus nerka*) and surveys of the abundance of presmolts have been carried out since 1981 as part of a program aimed at evaluating the productivity of the lake and its response to hatchery releases.

Early surveys served to establish that immigration of 0-group presmolts and emigration of smolts from the lake were complete by July. Studies of juvenile sockeye salmon throughout their range have shown that they exhibit distinct diel changes in distribution, and acoustic surveys are typically conducted at night to take advantage of the fact that they disperse throughout the midwater regions then.

The situation in Lake Tustamena is complicated by the fact that there is virtually no dark period during June and July, and that significant concentrations of fish exist in the upper 2.5 m of the water column even during the hours of darkness that exist in September (Thorne 1983). Near-field acoustic effects and vessel avoidance problems precluded sampling the region less than 2.5 m deep from a surface vessel. A system was therefore developed to examine the vertical distribution of the fish using up-looking transducers (Fig. 3-27) that could be deployed throughout the lake. Most up-looking deployments used a 70-kHz Simrad EY-M scientific echosounder with a wide angle

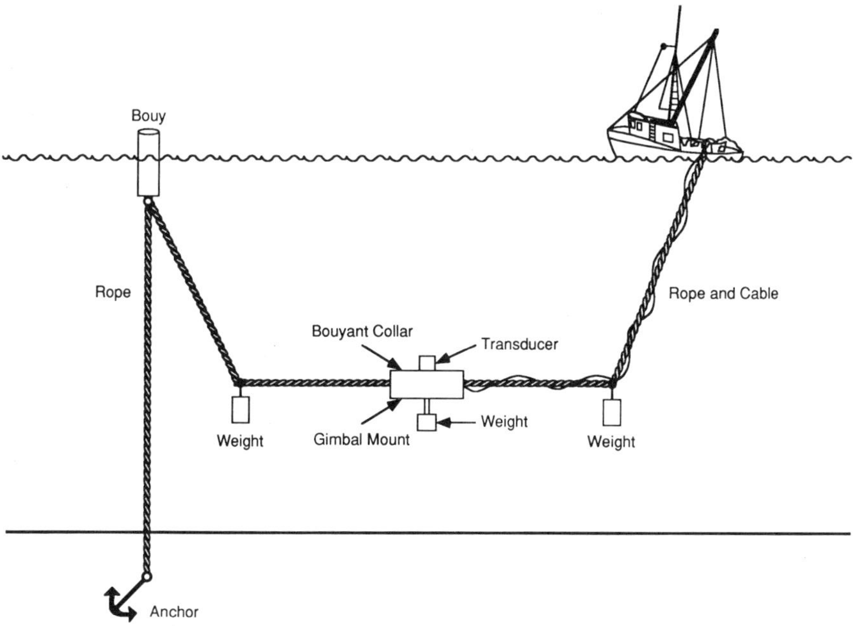

Figure 3-27. Up-looking acoustic system developed for surveys of Lake Tustumena. From Thorne and Thomas (1984).

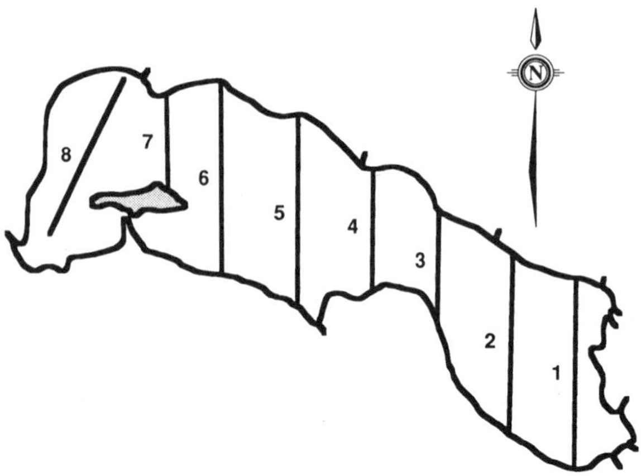

Figure 3-28. Transect plan for an acoustic survey of juvenile sockeye salmon in Lake Tustumena, Alaska. After Thorne and Thomas (1984).

(22°) transducer, and it was possible to detect fish targets within 0.3 m of the surface.

Data collected with the up-looking system were used to estimate the proportion of fish in the upper 2.5 m, and correct population estimates obtained by the down-looking survey for depths greater than 2.5 m. The results from surveys conducted over the years have shown that the corrections required are smallest if surveys are run at night, during September.

Transect locations for the survey vessel were determined by dividing the long axis of the lake into eight blocks. One transect was then randomly located within each block (Fig. 3-28), and run twice during each survey. Surveys were carried out using 7–9 m outboard vessels similar to the one shown in Figure 3-29, although a bow-mounted transducer was used to minimize avoidance

Figure 3-29. A small research vessel conducting an acoustic survey in a freshwater lake. Courtesy of BioSonics, Inc.

by near-surface fish. Each fully replicated survey required about 24 hours to complete.

The acoustic equipment employed during the down-looking surveys has varied over the years. In the early years of the study (Thorne 1983), an EK-120 (120-kHz) Simrad Scientific Sounder was used in conjunction with a 10° (full angle, 6 dB down) transducer, 40 log *R* TVG, and a pulse length of 0.6 ms. To minimize avoidance of the survey vessel by the fish, the transducer was mounted in a 2-foot Braincon V-fin and deployed from a bow-mounted boom. In later years a BioSonics 101 (420-kHz) echosounding system with a 12° beam width and a pulse length of 0.6 ms was used (Thorne and Thomas 1984).

All acoustic data were recorded on magnetic tape, and echo counting was carried out after the survey, using a storage oscilloscope. The threshold for counting was approximately 30 dB below the largest fish target amplitudes. Numbers of echoes from fish were counted in 10 minute increments along transects and in several depth intervals: 2.5–5.0, 5.0–9.5, 9.5–15.5, 15.5–27.5 m, and 27.5–38 m. Volume sampled in each depth interval was estimated using the duration in beam technique.

The echo counts and sampling volume measurements were used to calculate the density of fish (number/m^3) in each depth interval as the vessel proceeded over a given transect. These were then summed (Equation 3.4) across depth strata to determine the mean number of fish per square meter of surface for each run of each transect.

Identification of acoustic targets was carried out by a second vessel that accompanied the acoustic vessel and was equipped with a midwater trawl. The trawl had a 2.7 by 2.7 m mouth opening framed with PVC pipe. Mesh size decreased from 13.8 cm in the forward section to 0.6 cm in the rear, and a 0.3 cm liner was installed in the last 0.6 m. The trawl was towed at 3 m/sec, using 9-m bridles attached to each corner.

Most tows were directed at specific concentrations of fish identified by the acoustic vessel. An additional series of tows (about 10 out of 30 tows on a typical survey) were made while the acoustic boat was collecting data with the stationary, up-looking transducer.

Juvenile sockeye salmon dominated the trawl catches, making up 96.2–99.5% of the catches during the 1983 surveys (Thorne and Thomas 1984). The catches made for each run of each transect were subdivided into age 0 and age 1 groups, and used to apportion the acoustic density estimates (number/m^2) into age groups. These were then extrapolated to obtain population estimates for the area represented by a given transect, and summed to estimate the population estimate for the entire lake. On some occasions the availability of catch data made it necessary to combine the results for two transects, and in these situations, a weighted mean (weighted by transect length as in Equation 3.5) density estimate was used to calculate population size.

The population estimate for each age group, and its variance, was estimated from the mean and variance of the two replicate runs for each survey (see

Equations 1.3 and 1.4). These estimates did not include fish in the upper 2.5 m of the water column, and were adjusted upward using data from the up-looking system. The ratio of density in the 0–2.5 m/2.5–5 m depth intervals was used to do this, and ranged from 1.6:1 (July) to 1.0:1 (September) in 1983. The 95% confidence intervals for each survey ranged from ±18% of the estimated population size (all age groups combined) in July to ±40% in September during the 1983 surveys (Thorne and Thomas 1984). The limited nature of the up-looking data did not allow estimation of the variance associated with the correction for fish in the upper 2.5 m, and consequently the variances computed actually underestimate the true variance.

Applications of acoustic techniques in other sockeye salmon nursery lakes sometimes require the use of echo integration techniques in regions where fish densities are too high to allow single targets to be resolved. In these situations, it is often possible to count fish in those areas of the lake where densities are low, and use the results to scale the echo integrator output. Independent estimates of target strength and precise calibration of the acoustic system are not required with this approach, and it has been employed successfully in Cultus Lake, British Columbia (Nunnallee 1983) and in Lake Washington, WA (Thorne 1983). In the case of the Cultus Lake study, acoustic estimates of population size agreed closely (within 2.5%) with those obtained by direct counts of outmigrants at a sampling weir near the outlet of the lake. Some of the more recent acoustic surveys of Cultus Lake have used echo integration exclusively, employing dual-beam methodology to determine target strength (Burczynski and Johnson 1986).

Echo Integration

The stock of walleye pollock (*Theragra chalcogramma*) in Shelikof Strait, Gulf of Alaska, has been the target of annual surveys since 1980, when a major fishery began to exploit this stock.

A number of ancillary acoustic surveys have also been conducted throughout the Gulf of Alaska to locate and census other spawning stocks, and have shown that the Shelikof Strait spawning stock accounts for the bulk of the fish present in the western Gulf. Even when the Shelikof Strait stock was near its lowest level of abundance (an order of magnitude lower than in the early 1980s) an extensive survey of the central and western Gulf (137–165° W, Karp 1990) showed that other spawning stocks accounted for no more than 102,000 out of 484,000 mt (21%) detected. Spawning aggregations outside of Shelikof Strait may exist for short periods of time at locations separated by great distances, however, and pose a formidable sampling problem. This is complicated further by the fact that pollock are also present in very low densities along the edge of the continental shelf, and pollock echo sign cannot be differentiated from other sound scatterers in such cases. As a result, uncertainties continue to exist in understanding the relative importance of spawning stocks outside Shelikof Strait, and the interrelationships between these and the main spawning stock within the Strait.

A number of acoustic and ichthyoplankton surveys have shown that the spawning migration into Shelikof Strait takes place in the winter and early spring, with peak spawning occurring between the end of March and early April. During this period, the fish are localized in an area that can easily be surveyed within a week's time (Traynor et al. 1990). Although diel changes in aggregation pattern are observed, the fish generally remain in midwater and are available for acoustic assessment during both daylight and darkness. Survey operations are consequently carried out over a 24-hour period.

A bottom trawl survey conducted in March 1983 showed that while pollock were present in the area where they could not be fully detected by an acoustic survey, they represented only about 7% of the total biomass present (Nelson and Nunnallee 1985). However, more recent results (Karp 1990) indicate that the importance of the on-bottom component may vary with the age composition of the stock, and that coordinated bottom trawl-acoustic surveys may be required in the future.

To be certain that migrations into and out of the survey area (and their effect on biomass estimates) were well-understood, surveys in 1981–1986 consisted of 3–5 complete surveys during March–April (Nelson and Nunnallee 1985). Results from these early surveys showed that abundance was usually stable during March (Table 3-5), and cruises in recent years have consisted of two surveys. While a zig-zag survey track was followed during the early years, parallel transect designs (Fig. 3-30) are currently employed.

During the early years of the survey, acoustic data were collected using 38-kHz systems developed by the Applied Physics Laboratory, University of Washington, and BioSonics, Inc. The transducer was mounted in a towed body to minimize the effects of turbulence and rough sea conditions on survey results, and deployed at depths of 5–20 m (about 60 m behind the vessel). Echo integration was carried out in real time using a digital integrator interfaced to a computer, and both analog signal and integration data were stored on computer tape. Beginning in 1988, system electronics and computer facilities were housed in a portable cargo van that could be mounted on any survey vessel (Fig. 3-31). Calibration was initially carried out both before and after each cruise, using calibrated hydrophones. Use of standard calibration spheres (Fig. 3-18) was initiated in 1988.

In 1990, a Simrad EK 500 integration system was employed during the surveys. This system consisted of a 38-kHz split beam transducer, an echo-sounder-receiver, and a Sun workstation computer . The transducer was mounted on the RV Miller Freeman's centerboard, which could be lowered about 4 m below the keel. The Bergen Echo Integrator software (Fig. 3-11, Knudsen 1990) was used to process and digitally integrate the signals from the echo-sounder system, and data were stored on computer files and optical disks. Calibration was carried out both before and at selected intervals during the survey, moving to areas sheltered from the weather and using standard calibration spheres.

Depending on equipment limitations, echo integrator density estimates (kg/m^3) were computed at 1 to 10 minute intervals along survey transects, and at

Table 3-5. Pollock biomass estimates determined from Shelikof Strait Acoustic-Midwater Trawl Surveys

Year	Survey Number and Period	Biomass (t × 10[6])	95% Confidence Interval (t × 10[6])
1981[a]	1 March 3–15	4.38	2.92 to 5.84 (±33.3%)
	2 March 24–27	3.15	2.07 to 4.23 (±34.3%)
	3 April 4–10	3.06	2.02 to 4.08 (±33.3%)
1982		No survey	
1983[a]	1 March 6–15	2.46	1.54 to 3.40 (±37.7%)
	2 March 16–19	2.36	1.26 to 3.46 (±46.6%)
	3 April 6–13	0.82	0.57 to 1.07 (±30.5%)
1984[a]	1 March 3–9	2.03	1.43 to 2.64 (±29.9%)
	2 March 9–16	1.57	1.31 to 1.84 (±17.1%)
	3 March 16–18	1.90	1.06 to 2.75 (±44.2%)
	4 March 22–25	1.72	0.98 to 2.46 (±42.9%)
	5 April 1–7	1.66	1.19 to 2.13 (±28.3%)
1985[a]	1 February 21–28	0.65	0.47 to 0.82 (±27.3%)
	2 March 1–9	0.70	0.48 to 0.92 (±31.3%)
	3 March 14–20	0.77	0.55 to 0.98 (±27.8%)
	4 March 21–28	0.71	0.51 to 0.91 (±27.8%)
1986[a]	1 March 5–12	0.65	0.41 to 0.90 (±37.9%)
	2 March 13–21	0.60	0.39 to 0.80 (±34.5%)
	3 March 22–30	0.62	0.33 to 0.91 (±46.6%)

Annual Estimates

Year	Biomass (t × 10[6])	95% Confidence Interval	Source of Estimates
1981	3.77	2.86–4.67	Mean for surveys 1 and 2
1982		No survey	
1983	2.43	1.69–3.13	Mean for surveys 1 and 2
1984	1.84	1.21–2.47	Mean for surveys 1, 2, 3
1985	0.70	0.50–0.91	Mean for surveys 1, 2, 3
1986	0.62	0.37–0.87	Mean for surveys 1, 2, 3

[a]Estimates for 1981, 1983, and 1984 include very small amounts of age-1 fish, e.g., the biomass of age-1 fish in the "annual estimates" for these years did not exceed 0.024% of the total biomass. The 1985 and 1986 estimates include only fish age 2 and older.

Source: From Nelson and Nunnallee (1987).

1, 10, or 20 m contiguous depth intervals between the transducer and the bottom. Depending on weather conditions, integration could usually be carried out to within 3–5 m of the bottom. The limits for each survey's geographic coverage were partially determined by real-time observations of echo sign.

Target identification was carried out using midwater trawls. During 1990, a modified Northern Gold 1200 rope trawl equipped with 227-kg wingtip weights and 5 m^2 doors was employed. The net was constructed of ropes in

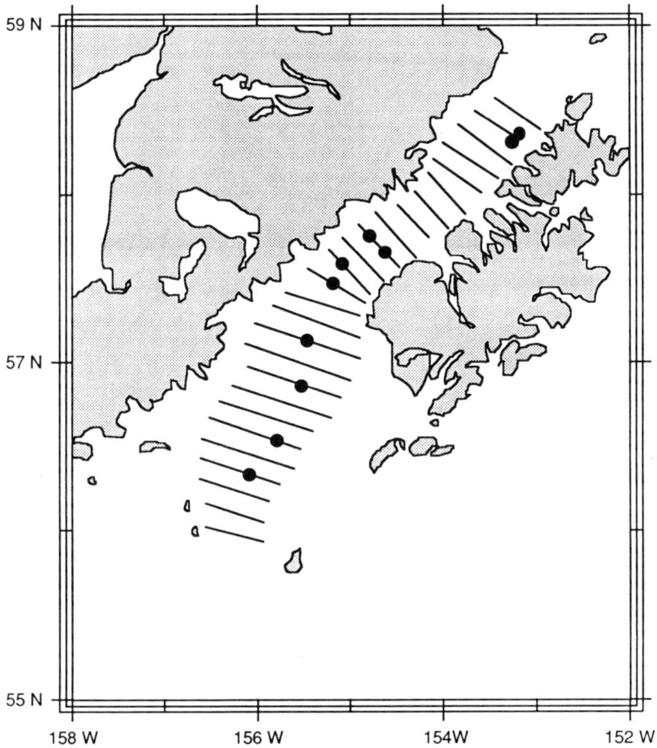

Figure 3-30. Transect plan for an acoustic survey of pollock in Shelikof Strait, Alaska. Solid dots show location of midwater trawl hauls. After Nunnallee and Williamson (1989).

the forward section and mesh sizes decreasing from 163 cm immediately behind the rope section to 8.9 cm in the cod end, and had a 3.2-cm liner inserted in the cod end. The vertical opening for this net usually ranged from 18 to 25 m at a towing speed of 3 knots.

Midwater trawls were equipped with acoustic mensuration systems, and their depth and vertical opening were monitored throughout each tow. All trawl hauls were made in order to identify the species composition of echo sign, and to provide data on the size and age composition of any pollock present. Fish near the bottom were sometimes sampled with demersal trawls.

All catches were processed using the sampling techniques described in Chapter 2 to obtain estimates of the species, size, and age composition of the catch. Pollock dominated both midwater and bottom trawl catches in Shelikof Strait. During 1980–1990, the proportion of pollock (by weight) in midwater trawl catches ranged from a high of 98% in 1981 to a low of 92% in 1989. Eulachon (*Thaleichthys pacificus*) or Pacific cod (*Gadus macrocephalus*) were usually second in importance, but their location in the water column and/or low target strength values were such that the integration data were never adjusted for their presence.

Figure 3-31. Typical configuration of a survey vessel during acoustic surveys by the Alaska Fisheries Science Center (NMFS/NOAA). Computers and electronic systems are located in the cargo van near the center of the vessel, and the transducer (visible at the water's surface) towed using an articulating crane. A midwater trawl is deployed from the hydraulic reel located near the stern. Photo courtesy National Marine Fisheries Service.

Echo returns showed that scattering layers were present at different depths, and midwater trawl data showed that the size and age composition of the pollock differed between layers. Area and depth strata were consequently defined on the basis of both integration and size/age data. The length frequency data for each trawl haul were weighted by the unscaled integration values obtained in adjacent areas (as described earlier in this chapter), then combined by strata.

Target strength data have been collected *in situ* during the Shelikof Strait surveys, using both dual beam and split beam technology. For each stratum, mean acoustic cross section ($\overline{\sigma}$) was estimated using the size composition data from the midwater trawl hauls and pollock TS–length data that had been collected on an opportunistic basis.

In recent years, the relationship $\text{TS} = 20 \log L - 66.0$ (Foote and Traynor 1988) has been used to obtain a target strength value for each length group. These target strength values were then converted to acoustic cross sections (σ), and weighted by the relative abundance of each size group in the trawl catches (Fig. 3-32) to estimate the mean acoustic cross section ($\overline{\sigma}$) for the population. The $\overline{\sigma}$ value was then used to scale the integrator output (volume backscattering or S_v) and estimate number m^3 for each depth stratum. Summing these values over depths (as in Equation 3.4), averaging them and

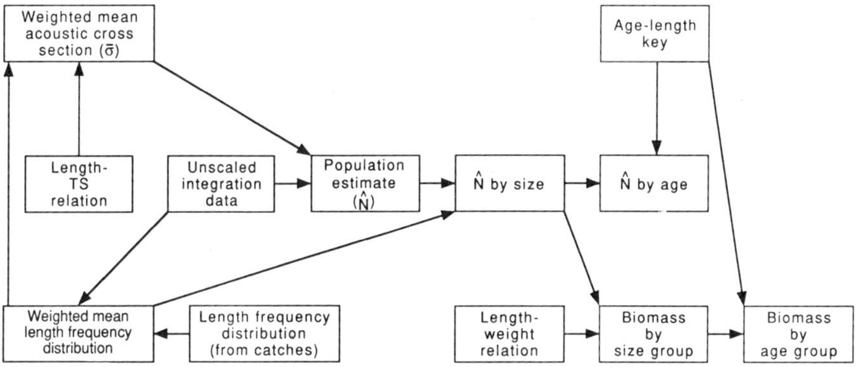

Figure 3-32. Estimation of population number and biomass by size and age group using catch and acoustic data.

extrapolating to the area covered by the stratum yielded population or biomass estimates by stratum, and for the survey area as a whole.

Variances for these estimates were obtained using Williamson's (1982) cluster sampling estimator. Annual biomass estimates were obtained by averaging the results from those surveys made during peak abundance (Table 3-5).

A more detailed analysis of the spatial distribution of pollock in 1987 (Sullivan 1991) indicated that the variances for the biomass estimates can be reduced substantially below the levels shown in Table 3-5, by adding additional depth strata to the survey design. His analysis did not cover the entire survey area, but yielded a CV of 9.9% when using a standard approach of treating each transect as a single observation, but using only a single depth stratum. By defining 12 10-m depth strata, the CV was reduced to 3.0%, and by applying spatial statistics (kriging) to depth-detrended data, it was reduced to 1.9%.

Length and age-specific biomass and population estimates were also calculated. In recent years this has been accomplished using the population estimate for each stratum, a length–weight relationship, acoustically weighted length frequency data, and an age–length key (Fig. 3-32). Since the length frequency data and age–length key (see Chapter 2) are obtained from the trawl sampling data, it must be assumed that trawling provides a representative sample of the population.

Bias in the estimates of biomass has been evaluated by comparing acoustic estimates with those made from ichthyoplankton surveys, and from catch–age analyses. The acoustic estimate for pollock 3 years and older in 1981 was 3.4 million mt while estimates from cohort analysis and ichthyoplankton surveys were 2.4 and 3.1 million mt, respectively (Kim and Gunderson 1989). The 95% confidence limits for the acoustic and cohort analysis estimates were ±24 and ±48%, respectively, but no variance estimate was made for the ichthyoplankton estimate.

The size of the stock in Shelikof Strait declined drastically during 1981–1990, due to the passing of a series of strong year classes through the fishery. The relative importance of the small stocks outside of the Strait has consequently increased. This has made it difficult to compare biomass estimates from acoustic surveys in Shelikof Strait with those from catch–age analyses for the Gulf of Alaska as a whole in recent years.

4

EGG AND LARVAL SURVEYS

INTRODUCTION

Ichthyoplankton sampling has played a central role in the development of marine biology and biological oceanography. Much of the earliest work was concerned with the identification and descriptive biology of eggs and larvae, however, and the first quantitative ichthyoplankton survey was conducted in the North Sea in 1895 (Hensen and Apstein 1897). The improvement of survey techniques has received considerable emphasis ever since, in order to fulfill the need for quantitative estimates of abundance, mortality, and advection rates of fish eggs and larvae.

Currently, a variety of fisheries oceanography projects are being undertaken in order to understand the processes that determine recruitment fluctuations in commercial fisheries, and quantitative ichthyoplankton surveys are an indispensable component of these studies. The primary objective of these studies is an understanding of the population dynamics of egg and larval stages, and the biotic and abiotic processes that determine recruitment. They are typically carried out in concert with a variety of related studies of the physical environment, availability of suitable food organisms, and predator abundance.

Early planktonologists recognized that it might be possible to use egg and larval surveys to deduce (back-calculate) the size of the spawning stock that produced them (Buchanan-Wollaston 1926). Obviously, this requires a knowledge of the spawning biology of the target species and the fecundity of adult females. It also requires the ability to identify the eggs or larvae of the target species, and an intimate knowledge of its early life history. The area and time of spawning must be well-defined, and the development, mortality, and advection rates of eggs or larvae determined. Nevertheless, this has been done successfully in many parts of the world. Egg and larval surveys provide population estimates that form the basis for the management of several anchovy, sardine, mackerel, and herring resources.

A number of different methods have evolved for using egg and larval surveys to assess fisheries resources (Table 4-1). These are usually dictated by the life history of the target species. For example, different sampling gear and techniques are required if we are surveying demersal eggs, pelagic eggs, or pelagic larvae.

Adult reproductive biology forces similar adjustments in methodology. Quite different approaches are required for determinate spawners, where the number of eggs to be released by a given female is fixed well in advance of the spawning season, as opposed to indeterminate spawners, which are continuously maturing new broods of eggs during the spawning season. While these characteristics force some catagorization of the different approaches that have been taken to population estimation using egg and larval surveys, it is important to realize that no "cookbook" of hard and fast rules exists. In

Table 4-1. Equations for population estimates using the annual egg production (AEP), daily egg production (DEP), and daily fecundity reduction (DFR) methods

Annual Egg Production Method	Daily Fecundity Reduction Method	Daily Egg Production Method
$\hat{P}_s = \frac{1}{\overline{F} r_s} E_a$ (4.4)	$\hat{P} = \frac{E_t}{R_t r}$ (4.5)	$\hat{P}_s = \frac{E_t}{\overline{F}_b S r_s}$ (4.6)
where $E_a = \int_{t=0}^{365} E_t d_t$ where $\hat{P}_s$ = size of spawning population E_a = total annual egg production in survey area $\overline{F}$ = mean annual fecundity per adult female r_s = proportion of adult females in spawning population (sex ratio) E_t = daily egg production (Equation 4.1)	where $\hat{P}$ = total population size R_t = daily reduction in fecundity per reproductively active female = $d(\overline{F}_t \overline{G}_t)/dt$ where $\overline{F}_t$ = mean fecundity per female with active ovaries (advanced yolked oocytes) on day t $\overline{G}_t$ = mean fraction of females with active ovaries on day t r = proportion of females in population	where $\overline{F}_b$ = mean batch fecundity = number of eggs released per day by a spawning female S = fraction of mature females spawning per day

fact, these methods are evolving rapidly, and will continue to do so in the future. In designing an egg or larval survey then, the assessment scientist should carefully evaluate the life history and distributional characteristics of the target species, and design the survey accordingly. It may be necessary to employ a combination of techniques in order to take full advantage of the opportunities presented by the target species. As we will see, a combination of egg survey techniques with acoustic surveys may prove advantageous in some cases.

REPRODUCTIVE BIOLOGY

Oocyte Maturation

Fish eggs undergo marked changes in size and character as they develop in the ovary. A small number of eggs first differentiate from the tiny (less than about 0.1 mm in diameter), transparent oocytes found in the oogonial nest, and begin to undergo vitellogenesis, growing in size as yolk is deposited within the cell. They become opaque, and develop a yellow, orange, or creamy coloration (depending on the species) due to the yolk deposits. Just before ovulation, the oocytes take up large quantities of fluid, or "hydrate" and once again become transparent. The oocytes can undergo an increase in volume of 2- to 3-fold during hydration. Spawning is usually imminent once ovulation has occurred, and the eggs can be readily extruded by light pressure on the sides of the body.

Some oocytes fail to complete this process, however, and undergo atresia prior to ovulation. The yolk within these oocytes is resorbed, and the oocyte and surrounding follicle undergo degeneration. In many species atresia is relatively uncommon, and in Dover sole (*Microstomus pacificus*) no more than about 1.5% of advanced oocytes are atretic (Hunter et al. 1992). In contrast, 10.4% of the yolk bearing oocytes of winter flounder (*Pseudopleuronectes americanus*) and about 5% of the oocytes produced by mackerel (*Scomber scombrus*) were found to be atretic (Tyler and Dunn 1976; Priede and Watson 1993). Atresia can significantly reduce potential fecundity in some species, and needs to be examined on a case by case basis.

Both prevalence (the proportion of the fish within a sample with ovaries showing signs of atresia) and intensity (proportion of atretic oocytes within an ovary) must be determined in order to estimate rates of atresia. Priede and Watson (1993) found that the incidence of atresia increased as mackerel approached the end of the spawning season, and that entire batches of oocytes were often aborted. The pattern of both prevalence and intensity of atresia suggested that atresia is a mechanism for adjusting egg output to available energy resources. Atresia near the end of the spawning season seems to be common in fishes with indeterminate fecundity, with maturing oocytes being resorbed.

The oocyte maturation and spawning process can be followed through gross examination of the gonads with the naked eye, and a number of classification schemes have been developed to standardize this. A system with five stages (Table 4-2) will often suffice. Some investigators add two or three additional stages to provide greater detail, while others (e.g., Hunter et al. 1992) use a four-stage classification system. The subjectivity involved often limits the utility of any of these systems however, since they represent attempts to impose discrete categories on what is a continuous process. It is usually quite difficult to differentiate immature (juvenile) fish from adult fish in the resting stage, and any studies to determine size and age at maturity must be undertaken near the spawning season.

The size distributions of the oocytes can be obtained by separating them from the surrounding ovarian tissue, and measuring their diameters under a microscope. A number of modes will usually be evident (Fig. 4-1). These modes reflect the development process, corresponding to undifferentiated (immature), maturing, or hydrated oocytes. If followed over the course of the year, these size distributions reveal whether the species in question is a determinate or indeterminate spawner.

A determinate spawner will have no more than three major modes, corresponding to immature, maturing, and hydrated oocytes. If all oocytes mature as a single group, and are spawned simultaneously ("total" spawning), there will be only two modes, corresponding to undifferentiated and mature oocytes. However, some vitellogenesis continues during the spawning season in many determinate spawners, and they hydrate small batches of their mature eggs over a prolonged period. For example, Dover sole are determinate spawners, maturing only a single brood of eggs each year, but spawn about 9 batches over a 5- to 6-month spawning period before all of these eggs have been released (Hunter et al. 1992). Atlantic cod spawn 17–19 batches of eggs over a period of about 50–60 days (Kjesbu 1989). This may represent either a mechanism for a fish with limited body capacity to posess a high fecundity,

Table 4-2. Ovarian development and maturity stages

Maturity Code	Stage	Description of Ovary
I	Immature (juvenile)	Small and translucent
II	Resting	Firm and opaque; oocytes not visible to the naked eye
III	Maturing	Orange, yellow or cream-colored; oocytes filling with yolk, and may be visible to the naked eye
IV	Ripe	Contains eggs that are large, translucent, and can often be extruded by pressure on side of body
V	Spent	Baggy and bloodshot, containing only a few residual eggs

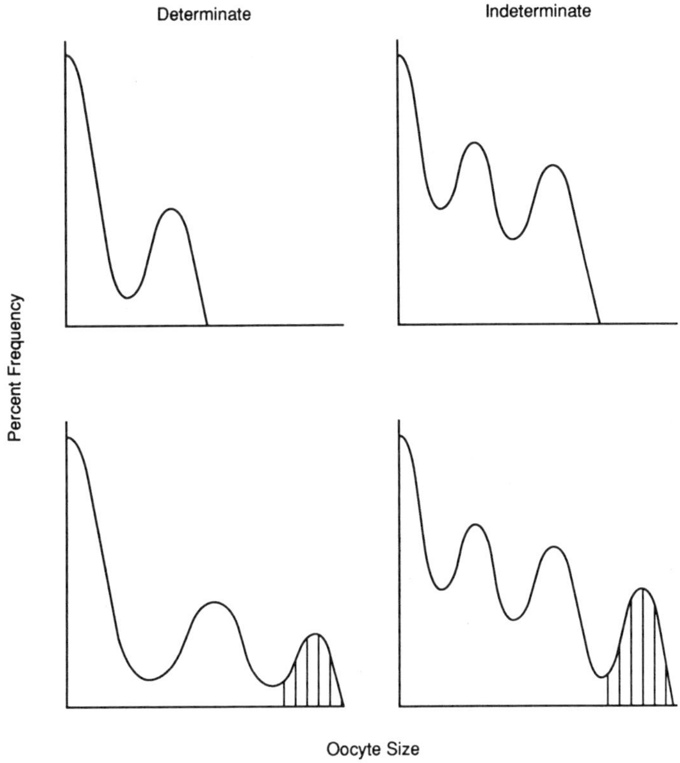

Figure 4-1. Oocyte size distribution during maturation (upper panel) and spawning (lower panel) for determinate versus indeterminate spawners. Shaded areas indicate the mode corresponding to hydrated oocytes.

or for spreading the risk in such a way that all offspring are not subject to mortality during a brief period of adverse conditions. Total annual fecundity for determinate spawners can be estimated by counting the number of oocytes present in ovaries that have been collected at a time when maturation is nearly complete, yet no spawning has taken place. Some of these oocytes may eventually be resorbed, however, and it is important to estimate rates of atresia and make appropriate reductions in the fecundity estimates. Rates of atresia can be estimated by obtaining representative samples of maturing females, and examining histological preparations of ovarian sections (Hunter and Macewicz 1985a; Priede and Watson 1993).

In contrast, indeterminate spawners differentiate and mature several broods of eggs during the year. Size frequency plots for these fish can consequently show more than three modes, corresponding to different broods. Typically, the oocytes can show an almost continuous range in sizes (Fig. 4-1). No large gaps exist between size classes of oocytes except for one between hydrated and advanced yolked oocytes. There will rarely be evidence of more than a few developing broods in a single fish, and it is impossible to determine just how many broods will eventually be produced without repeated observations on the same fish. It has been estimated that northern anchovy (*Engraulis mordax*) produce an average of about 20 broods of eggs annually off California (Hunter and Leong 1981), even though no more than a few of these are present in the ovary at a given time. The number and size of eggs produced annually by indeterminate spawners can be quite flexible. Laboratory studies of the Japanese anchovy (*Engraulis japonica*) showed that a reduction in the food ration below the level required for maintenance resulted in a prolongation of the interspawning interval, a reduction in batch fecundity, and a reduction in mean egg size (Tsuruta and Hirose 1989). In wild populations the interspawning interval, batch fecundity and egg size can vary widely, both between years and over the course of a single spawning season (Hunter and Leong 1981; Alheit 1988, 1989).

Indeterminate spawners are sometimes referred to as "batch," "serial," "partial," "heterochronal," or "multiple" spawners, terms that fail to adequately distinguish them from those determinate spawners that also release their eggs in batches, rather than all at once. Species such as herring, which release all of their eggs in a single spawning event, are termed "total" or "isochronal" spawners.

Histological examination of the ovary provides a more detailed picture of the maturation process, and should always be undertaken in order to determine the frequency of occurrence for atretic follicles. Histological studies are also valuable in validating assumptions as to whether a species is a determinate spawner.

In the case of indeterminate spawners, histological examination of the ovaries provides an opportunity to estimate spawning frequency (percent spawning per day) and makes it possible to relate the number of eggs in the sea to the size of the spawning stock that produced them. Four distinct oocyte maturity stages are associated with spawning:

1. The migratory nucleus stage for yolked oocytes, when the nucleus migrates to the animal pole. This stage indicates that hydration is imminent.
2. The hydrated oocyte stage, which indicates that ovulation is imminent. This stage is the most ephemeral of all oocyte classes, and does not occur earlier than 12 hours before spawning in northern anchovy (Hunter and Macewicz 1985b).
3. The early (0–5 hours old in northern anchovy) postovulatory follicle stage.
4. The late (19–28 hours old) postovulatory follicle stage.

The follicle surrounding the oocyte ruptures and collapses on ovulation, but an identifiable postovulatory follicle still remains. This follicle deteriorates rapidly and is resorbed. Within 48 hours of ovulation, postovulatory follicles of anchovies can no longer be differentiated from atretic follicles (Hunter and Macewicz 1985b). The rate of resorption is even more rapid in the case of tropical skipjack tuna (*Katsuwonus pelamis*), and postovulatory follicles are identifiable for only 24 hours (Hunter et al. 1986).

To differentiate and determine the age of postovulatory follicles, it is desirable to hold mature specimens of the target species in the laboratory, induce them to spawn, then sacrifice them at regular intervals. Histological preparations from such a series enabled Hunter and Goldberg (1980) to develop histological criteria for the differentiation of age 0-day and age 1-day postovulatory follicles in anchovy.

If spawning is synchronous in the wild, occurring within only a few hours out of the day, laboratory rearing can be avoided by sampling spawning schools over a 24-hour period in order to obtain the required histological series (Alheit et al. 1984). In principle, any of the four stages associated with spawning can be used to determine spawning frequency, provided their duration is known.

Spawning

Spawning migrations and behavior have an important bearing on how useful egg and larval surveys will be in assessing the population for a given species. Almost all freshwater species and many marine species (notably herring and capelin) have demersal eggs, either attached to the substrate, buried within it, or drifting loosely over the bottom. The problems associated with sampling demersal eggs are such that quantitative surveys are uncommon. Pacific herring attach their eggs to intertidal or subtidal vegetation, however, and can be surveyed by SCUBA divers using quadrat sampling techniques.

Pelagic eggs offer a more tractable sampling problem, since they can be collected with the quantitative plankton nets commonly employed in biological oceanography. Surveys of species such as anchovies, sardines, walleye pollock, mackerel, North Sea plaice, and Atlantic cod, all of which have pelagic eggs, have been successfully undertaken, and the results have been used to estimate population abundance.

Most species undergo well-defined migrations between spawning and feeding areas. In many marine species these migrations enable the species to exist within a given current system, yet still have passively drifting egg and larval stages. Adult migrations are often contranatant (upstream) when moving toward the spawning ground, while the drift of eggs and larvae is denatant (Harden-Jones 1968). In other species, spawning takes place in estuaries, mesoscale eddies, or frontal systems where the hydrography is such that eggs and larvae will be retained in the spawning area (Sinclair 1988).

Spawning must occur within a well-defined area and depth range in order to employ the methods described in this chapter, and migrations to and from this area must be thoroughly understood. A central assumption is that all eggs produced by the target species will be encompassed within the survey area, and advection of eggs into and out of this area is negligible.

Spawning can occur at considerable depth for some species, and it is imperative that sampling encompasses the entire depth range that the eggs are found in. Off California, Dover sole (*Microstomus pacificus*) spawn near the ocean floor at depths of 600–1500 m, and the eggs rise to the surface after fertilization. Sampling Dover sole eggs requires considerable ship time and depth-instrumented sampling gear as a result (Lo et al. 1993). In contrast, the maximum depth that needs to be sampled for anchovy eggs is typically 210 m (Smith and Richardson 1977).

The spawning period is well defined for most temperate species and often occurs with surprising regularity. A total of 12 egg surveys for North Sea plaice in the Southern Bight have been carried out since 1911, and have shown that peak spawning occurs on 19 January, with a standard deviation of less than a week (Cushing 1982). Arctic cod show a similar regularity (Fig. 4-2), the median date of spawning at Lofoten, Norway showing a standard deviation of only 2.2 days during 1968–1972 (Pedersen 1984). Regularity in the occurrence of the peak spawning period is an important consideration, since the accuracy of any of the techniques covered in this chapter is dependent on adequate survey coverage during peak spawning.

The duration of the spawning period is also an important consideration, particularly in the case of the annual egg production method (Table 4-1), where surveys must be repeated over the entire course of the spawning period. This obviously becomes impractical when spawning occurs over a prolonged period, as is commonly the case for many tropical and subtropical species. In the case of northern anchovy, some spawning takes place year-round off California, although little occurs in November and December (Parrish et al. 1986). A contrasting situation is found for walleye pollock in Shelikof Strait, Alaska, where almost all spawning takes place over a period of 2 months (Kim and Gunderson 1989).

Egg Development

Morphogenesis of fertilized eggs during the incubation period follows a predictable series of stages in fish (Blaxter 1969) , and forms the basis for "staging" eggs as to developmental status (see Table 4-3 and Fig. 4-3 for example). By spawning fish in the laboratory and rearing the eggs under carefully controlled temperature and salinity conditions, data on the duration of the different egg stages can be obtained (Fig. 4-4). This information can then be used to convert data on egg stages to egg–age data, and to estimate development and mortality rates in the field. These rates must be known in order to obtain accurate estimates of population size.

Egg development rates increase exponentially with temperature for most species (Blaxter 1969), and it is necessary to conduct laboratory rearing experiments over an appropriate temperature range, then relate these results

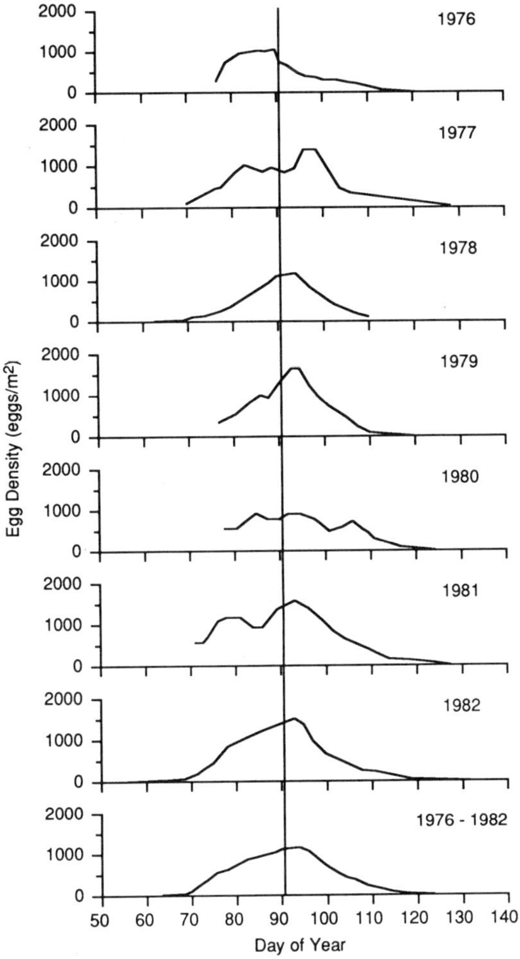

Figure 4-2. Spawning curves for cod spawning off Lofoten, Norway during 1976–1982, obtained through direct sampling on the spawning grounds. The vertical line corresponds to 1 April. After Pedersen (1984) based on data from Solemdal (1982).

to the the thermal history of field-caught eggs in order to age the eggs accurately. This is no easy matter for many species.

Specific gravity can change as an egg develops, and if the egg is pelagic, its location in the water column will fluctuate accordingly. The trajectory followed by developing eggs can be estimated by determining their specific gravity at different developmental stages, and correlating these results with water density–depth profiles. The specific gravity of walleye pollock eggs and larvae has been determined by placing them in density gradient tubes at different developmental stages (Kendall and Kim 1989). Specific gravity observations were then used in conjunction with seawater density observations (from conductivity–temperature–depth data) to reconstruct the trajectory

Table 4-3. Classification criteria for anchovy eggs[a]

Stage I: Cell division has not yet begun. The cytoplasm of the single cell appears as a clear hemisphere at one pole, easily differentiated from the yolk mass which is divided into granules. The cytoplasm may be displaced to other locations around the periphery of the yolk mass, but there is usually some accumulation at one pole, which allows the stage to be identified.

Stage II: This begins with the division of the single cell into two cells or blastomeres. The division is first noticeable when a furrow develops in the middle of the cytoplasmic cap. Small bubble-like structures (probably artifacts) are often visible along the furrow and help identify it. The next cleavage plane is at right angles to the first, and subsequent synchronous divisions in both meridional and latitudinal planes produce a hemispherical mound of cells, termed the blastodisc. After the 5th or 6th division, the blastodisc has a berry-like appearance, the so-called "mulberry stage," and with subsequent divisions the blastomeres become increasingly smaller and more difficult to distinguish individually. During a certain phase of the early divisions the blastomeres are about the same size as the yolk granules. If the blastodisc and yolk mass become disrupted during collection or preservation, the blastomeres may become distributed among the yolk granules. They may be distinguished from one another since they have different refractive indices and the blastomeres appear darker when viewed with transmitted light.

Stage III: This stage begins with the appearance of the segmentation cavity. The segmentation cavity of teleost eggs is the space formed between the blastodisc and the yolk mass during late cleavage. In most anchovy eggs the blastodisc is shrunken and somewhat cup-shaped and consequently the segmentation cavity, which is a delicate structure, is obliterated. We have found it preferable to define the beginning of Stage III on the basis of the appearance of the blastoderm, i.e., when it has the appearance of tissue rather than of a collection of individual cells. This stage marks the beginning of gastrulation. The margin of the blastodisc becomes slightly thickened and is termed the germ ring. At one region of the germ ring the thickening extends inward to form the embryonic shield, which defines the future axis of the embryo. Gastrulation proceeds by further proliferation and downward movement of cells in the region of the germ ring by a process known as epiboly. Simultaneously, proliferation and inward migration (emboly) of cells from the margin of the embryonic shield produce the organ-forming cell layers of the primordial embryo. At the end of Stage III the germ ring is one-third down the yolk mass and the bilateral nature of the primordial embryo is apparent.

Stage IV: At the beginning of this stage the germ ring has enclosed one-third of the yolk mass and the embryo is beginning to form along the medial region of the embryonic shield. At the end of this stage, defined by the germ ring enveloping two-thirds of the yolk, the head region of the embryo is becoming apparent.

Stage V: This stage begins with the germ ring two-thirds down the yolk and ends with the closure of the blastopore and the complete enclosure of the yolk by the cellular sheath of the embryo. The stage is characterized by rapid differentiation resulting in the formation of several somites in the midregion of the embryonic axis, development of the notochord which can be seen from a dorsal viewpoint, and differentiation of the optic vesicles from the brain.

Stage VI: This stage begins with the closure of the blastopore and ends when the tail starts to separate from the yolk mass. The embryonic sheath of cells is extremely thin and, in some samples, it may be difficult to determine the point of blastopore closure. In these cases the event can be estimated from the position of the caudal terminus of the embryonic axis, since it grows toward the pole with the edge of the cellular sheath. Initially the caudal region lies flat against the polar region of the yolk, then gradually thickens and becomes more rounded at the tip until it is clearly separate from the yolk. During this stage the somites are apparent along the entire body axis (except at the caudal portion), the lens primordium appears in the eye, and the regions of the brain begin to differentiate.

Stage VII: At the beginning of this stage the tip of the tail (free from the yolk) is broadly rounded, then begins to narrow as it elongates. The notochord extends almost to the tip and the finfold is just becoming visible. At the end of this stage the length of the free tail is one-half the length of the head. For this purpose, head length is considered the distance from the tip of the snout to the back of the cerebellum (see Fig. 4-3bH). Relative tail length is the criterion for each remaining stage.

Stage VIII: This stage begins when the free tail length is greater than one-half the head length and ends when tail length equals head length. The tail becomes pointed during this stage and begins to bend away from the axis of the body, to the right or left side. The curvature of the tail generally increases with development, but is subject to individual variability (Fig. 4-3b). Judgment is required in compensating for curvature in estimating relative tail length, although accuracy and precision increase rapidly with practice.

Stage IX: This stage begins with the tail extending one quarter the length of the yolk sac and ends when it reaches one-half the yolk sac length. The gut is now apparent along the ventral surface of the tail, and its terminal section passes through the finfold which is now considerably wider than in the previous stage. The pectoral fin buds appear as lateral thickenings as do the otic vesicles.

Stage X: This stage starts when the tail is one-half the length of the yolk sac and ends when it reaches three-quarters of the yolk sac length.

Stage XI: This is the final stage before hatching and is defined by a tail length greater than three-quarters of the length of the yolk sac.

---

[a]See Figure 4-3 for examples.

Source: Moser and Ahlstrom (1985).

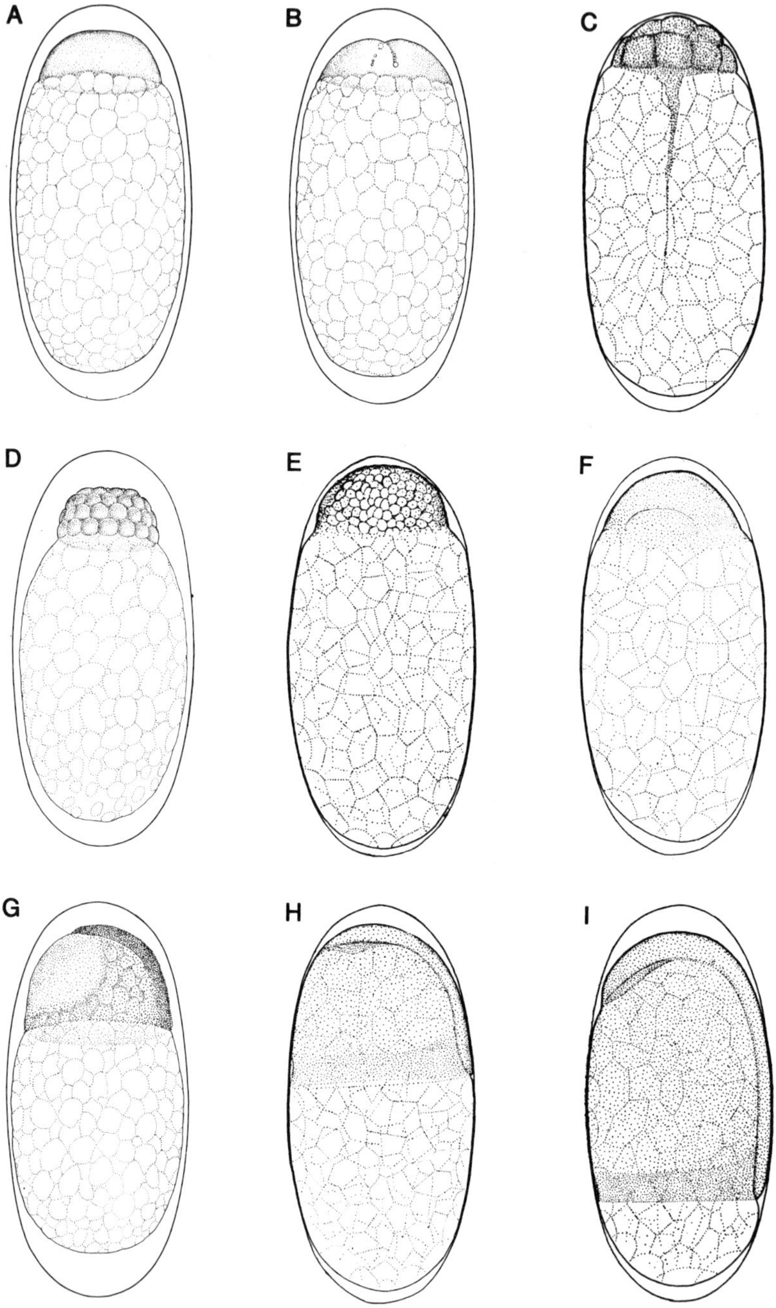

Figure 4-3(a). Stages of northern anchovy eggs. (A) Stage I, (B) Stage II (2 cells), (C) Stage II (16 cells), (D) Stage II ("Mulberry"), (E) Stage II (late), (F) Stage III (mid), (G) Stage III (late), (H) Stage IV (mid), (I) Stage V (mid). Original illustrations of A, B, D, G by G. Mattson; C, E, F, H, I from Bolin (1936). From Moser and Ahlstrom (1985).

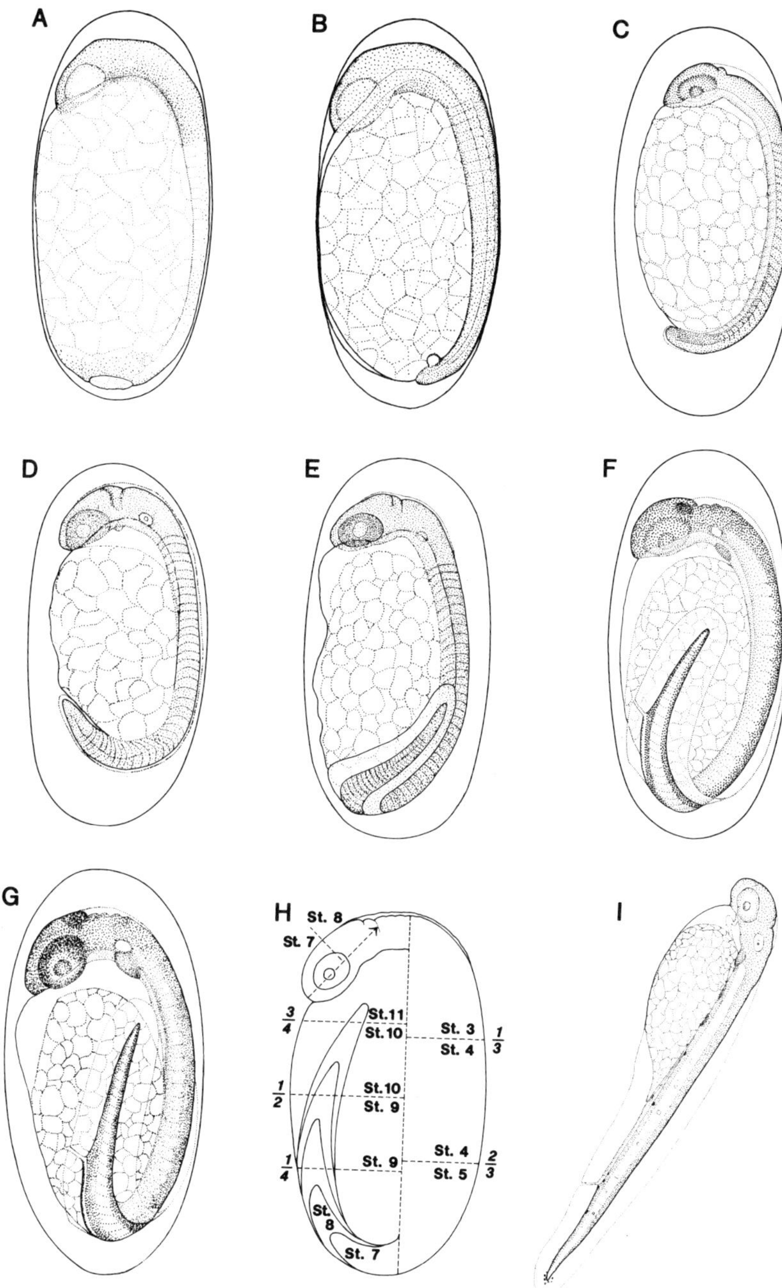

Figure 4-3(b). Stages of northern anchovy eggs and larvae. (A) Stage V (late), (B) Stage VI, (C) Stage VII, (D) Stage VIII, (E) Stage IX, (F) Stage X, (G) Stage XI, (H) diagram showing relationship between epiboly and tail length to stage; right side indicates fraction of yolk mass covered by the blastoderm in Stages III-V; left side shows tail length/head length proportions that define Stages VII and VIII and tail length/yolk mass length proportions that define stages IX-XI, (I) newly hatched larva. Original illustrations of A, C, and F by G. Mattson, B from Bolin (1936); D and E by H. Orr; G by G. Moser; H by B. Sumida; I from Kramer and Ahlstrom (1968). From Moser and Ahlstrom (1985).

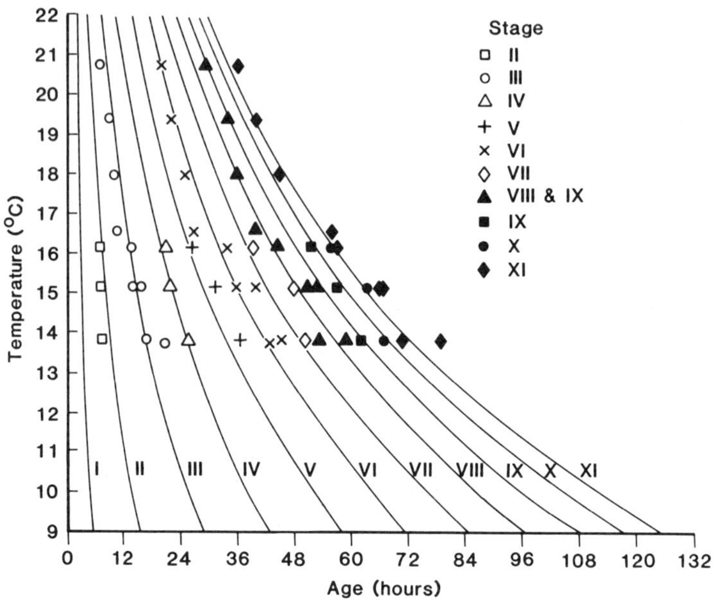

Figure 4-4. Observed (various symbols) and predicted ages (time to the *i*th stage, fitted curves) as a function of temperature (°C). Based on egg development data collected during laboratory experiments. From Lo (1985).

followed by eggs and larvae during the developmental process (Fig. 4-5). This trajectory was then used to determine the thermal history of developing eggs, and to estimate duration of developmental stages, ages, and mortality rates.

Data on the geographical and temporal distribution of eggs of different ages also allow inferences to be made as to the location and time of day that spawning is most intense. Rates of advection out of the spawning area can be estimated from these data as well (Koslow et al. 1985; Kendall and Kim 1989).

Larval Development and Gear Avoidance

The larval stage begins with hatching, when the larva leaves the egg shell, and ends with metamorphosis. Newly hatched larvae of many species have unpigmented and presumably nonfunctioning eyes, and it seems probable that the eye at first feeding is often capable of only coarse movement perception (Blaxter 1969). Generally the youngest larvae have a system of free neuromasts, which probably enable them to respond to water currents and disturbances. Little is known about the development of the inner ear, although otoliths are present at hatching and are used for aging larvae. Probably all larvae have a labyrinth that is sufficiently well developed to give equilibrium perception soon after hatching. It is much less certain whether they hear using the saccular or utricular maculae (Blaxter 1986).

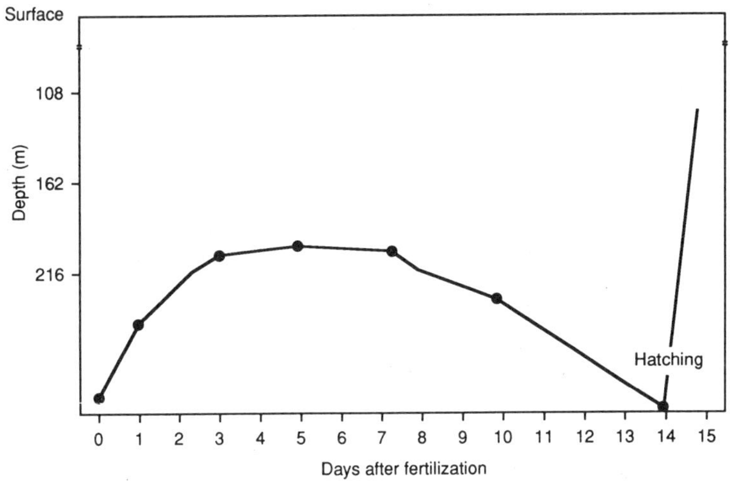

Figure 4-5. Projected vertical movement of walleye pollock eggs with development. After Kendall and Kim (1989).

Larvae undergo a series of important morphological changes as they grow, including development of rods in the retina of the eye (associated with light–dark adaptation), and lateral line development. Changes in the complexity of the eye during development lead to increased visual capabilities, and both reactive distance and search volume increase significantly with larval length (Miller et al. 1988). Larval burst and average swimming speeds increase as a linear function of length, further enhancing their ability to elude sampling gear. By 6–8 mm, larvae of many species are able to evade gear such as high speed Miller samplers (10 cm diameter, Noble 1970) and low-speed 1-m plankton nets (Murphy and Clutter 1972) during the day. By about 20 mm even gears with relatively large sampling aperatures are ineffective.

In herring, development of the lateral line, rods in the retina, and gas filling of the auditory bullae (associated with startle responses to sound) is not completed until metamorphosis (approximately 40 mm), although they all commence development at approximately 25 mm. During the day herring larvae are able to successfully evade a midwater frame trawl with a mouth opening of 5 m^2, towed at 2 m/sec, when about 25 mm long (Heath and Dunn 1990). Daytime experiments with anchovy (*Stolephorus purpureus*) indicated that some larvae were able to avoid purse seines with an effective sampling area of about 350 m^2 when about 20 mm long (Murphy and Clutter 1972).

Day–night comparisons have shown that visual cues are responsible for a major portion of larval avoidance behavior, and avoidance can be significantly reduced by sampling at night. However, other cues (vessel noise, noise or pressure fields created by the sampling gear etc.) are also important, since some avoidance still occurs at night (Murphy and Clutter 1972; Brander and Thompson 1989). Avoidance of sampling gear will probably vary with ambient water temperatures (which affect swimming speeds) and light conditions, and has been shown to vary from year to year (Noble 1970).

Problems Posed in Population Estimation

It often becomes difficult or impossible to quantitatively sample the larvae of many species beyond a length of about 10 mm, due to their ability to avoid standard icthyoplankton sampling gear. The development of more effective sampling equipment could improve this situation in the future, but significant problems would still remain if larval surveys are to be used to estimate adult abundance. This is due to the fact that it is necessary to relate larval abundance during a survey (or series of surveys) to the number of eggs that produced these larvae. Egg mortality rates during incubation and larval mortality rates during the period preceding the survey must all be known in order to do this properly.

Age determination of larvae can be accomplished by analysis of length frequency modes, or by counting daily growth rings laid down on the otolith. The resulting data on age composition allow growth and mortality rates to be estimated, and are required in order to back-calculate population size at hatching. Estimates of egg mortality during the incubation period must still be obtained, however, making it necessary to carry out further research on the egg stage. The logistics of sampling larvae quantitatively and making accurate estimates of the appropriate growth and mortality rates are formidable, and usually preclude accurate estimates of adult population size from data on larval abundance (Hauser and Sissenwine 1991; Heath 1993).

PELAGIC EGG SURVEYS

Survey Design

The first step in designing an ichthyoplankton survey is to determine the geographic limits of the survey area in such a way as to be certain that all of the eggs produced by the target population are present during the survey. Since the spawning area frequently changes from year to year, it is often necessary to take an adaptive approach to delineating the survey area. Egg abundance is monitored at each station, and sampling along a transect discontinued when no eggs are found in two to three adjacent stations. Post-stratifying the survey area by defining boundaries beyond which all hauls yielded zero catches is an effective way to treat the results of such a survey, and results in a negligable bias when the number of stations is large (Chapter 1).

The eggs present should constitute a closed population, with no significant gains or losses due to advection. Some preliminary work should be carried out with discrete depth samplers (e.g., Clarke-Bumpus samplers, Tucker or MOCNESS trawls; Snyder 1983) in order to determine the depth range that must be sampled.

Ichthyoplankton surveys typically follow a systematic grid design (Fig. 4-6). Several strata are usually developed on the basis of the egg abundance anticipated, with higher station densities being allocated to areas of high abundance. The density observation at each station is usually treated as a random variable, and the methods outlined in Chapter 1 used to estimate both the number of eggs present in the survey area and its variance. In some instances ichthyoplankton data have been poststratified by contouring, and as was the case with acoustics this is usually a poor idea due to the problems associated

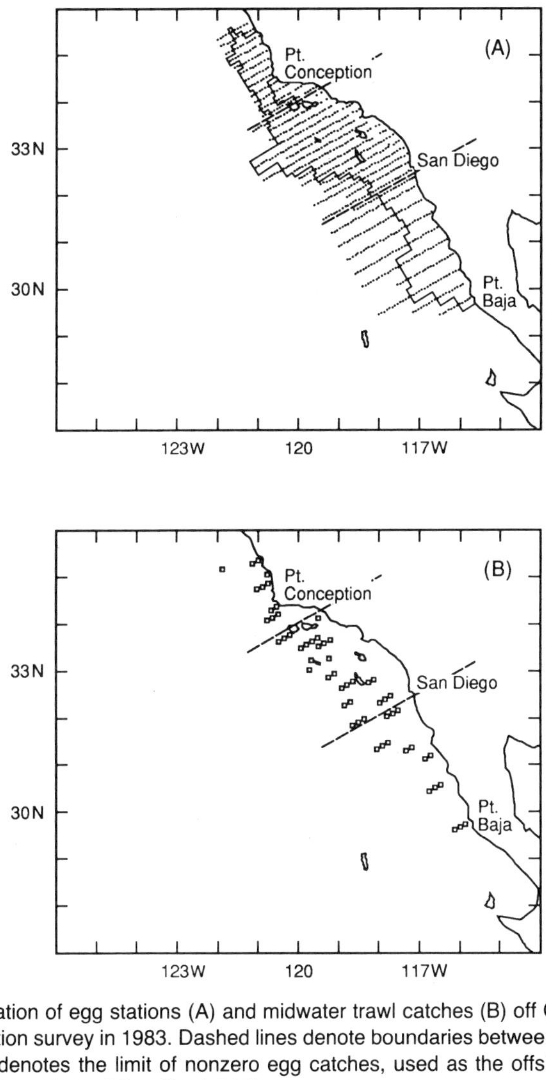

Figure 4-6. Location of egg stations (A) and midwater trawl catches (B) off California during a daily egg production survey in 1983. Dashed lines denote boundaries between latitudinal strata. Solid line in (A) denotes the limit of nonzero egg catches, used as the offshore boundary for all strata. After Picquelle and Stauffer (1985).

with variance estimation (see Chapter 3, Survey Design and Analysis of Acoustic Data).

Data Requirements

While there are three principal approaches to estimating population size from egg abundance data (Table 4-1), all three require an estimate (or several estimates) of the number of eggs produced per day ($\hat{E}_t$) during a given survey. This requires information on egg mortality and incubation rates (stage durations) as well as egg abundance, and is determined from

$$\hat{E}_t = (\hat{E}_{ij}/d_{ij})e^{Z_{ij}t_{ij}} \tag{4.1}$$

where $\hat{E}_{ij}$ = number of stage i eggs in the survey area during survey j
d_{ij} = duration (days) of egg stage i during survey j
Z_{ij} = instantaneous rate of daily mortality for egg stage i during survey j
t_{ij} = time (days) elapsed between fertilization and midpoint of stage i during survey j (age of egg)

The equation for estimating ($\hat{E}_{ij}$) from a stratified random survey (a modification of Equation 1.1) is

$$\hat{E}_{ij} = \sum_{k=1}^{h} A_k \overline{D}_{ijk} \tag{4.2}$$

where A_k = area within stratum k (m^2)
$\overline{D}_{ijk}$ = mean number of stage i eggs per m^2 in stratum k during survey j, based on n_k samples
h = number of strata

The sampling gear employed must be able to provide quantitative estimates of the number of stage i eggs per m^3 at each station. This is then multiplied by the depth of the water column sampled (R_m) to estimate the mean number of eggs per m^2 for each stratum

$$\overline{D}_{ijk} = \frac{1}{n_k} \sum_{m=1}^{n_k} C_{ijkm}(R_m)/\pi r^2 l_m \tag{4.3}$$

where C_{ijkm} = number of stage i eggs caught at station m
R_m = depth (m) of water column sampled at station m (maximum depth of sampler)
r = radius (m) of mouth of sampler
l_m = effective length of tow path (m) at station m, obtained from a calibrated flowmeter

Sampling Gear

While a wide variety of sampling gears are available for ichthyoplankton, many of these were developed to sample larvae capable of avoiding the gear. The simplest vertically deployed plankton nets (bridled circular plankton nets of various diameters, Juday nets etc.) will suffice for sampling eggs, however. Nevertheless, care must be taken to design, instrument, and deploy them properly when quantitative results are required. Figure 4-7 shows an example

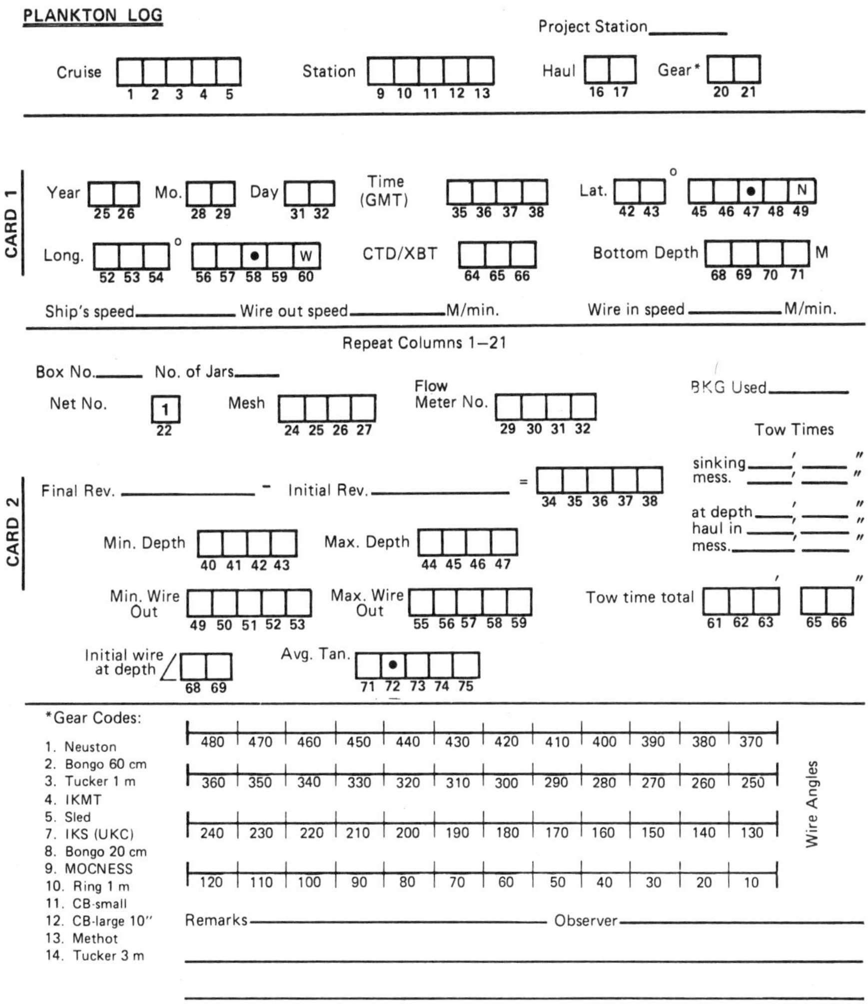
PLANKTON LOG

Project Station________

Cruise (1 2 3 4 5) Station (9 10 11 12 13) Haul (16 17) Gear* (20 21)

CARD 1

Year (25 26) Mo. (28 29) Day (31 32) Time (GMT) (35 36 37 38) Lat. (42 43) ° (45 46 47 • 48 49 N)

Long. (52 53 54) ° (56 57 58 • 59 60 W) CTD/XBT (64 65 66) Bottom Depth (68 69 70 71) M

Ship's speed________ Wire out speed________M/min. Wire in speed________M/min.

Repeat Columns 1–21

Box No.____ No. of Jars____

Net No. 1 (22) Mesh (24 25 26 27) Flow Meter No. (29 30 31 32) BKG Used________

CARD 2

Final Rev. ________ – Initial Rev.________ = (34 35 36 37 38)

Tow Times: sinking ___' ___" mess. ___' ___" at depth ___' ___" haul in ___' ___" mess. ___' ___"

Min. Depth (40 41 42 43) Max. Depth (44 45 46 47)

Min. Wire Out (49 50 51 52 53) Max. Wire Out (55 56 57 58 59) Tow time total (61 62 63)' (65 66)"

Initial wire at depth ∠ (68 69) Avg. Tan. (71 • 72 73 74 75)

*Gear Codes:
1. Neuston
2. Bongo 60 cm
3. Tucker 1 m
4. IKMT
5. Sled
7. IKS (UKC)
8. Bongo 20 cm
9. MOCNESS
10. Ring 1 m
11. CB-small
12. CB-large 10"
13. Methot
14. Tucker 3 m

Wire Angles:

480	470	460	450	440	430	420	410	400	390	380	370
360	350	340	330	320	310	300	290	280	270	260	250
240	230	220	210	200	190	180	170	160	150	140	130
120	110	100	90	80	70	60	50	40	30	20	10

Remarks________ Observer________

Figure 4-7. Example of a form used to record data on the sampling gear, location, flowmeter revolutions, wire angles, and other aspects of gear deployment during an ichthyoplankton survey. Numbers above blocks indicate column to be used for data entry. Courtesy of A. Kendall, National Marine Fisheries Service.

of the type of information that must be collected during a typical plankton haul.

A calibrated flowmeter should be placed in the net in such a way that it will provide an accurate measure of the effective distance traveled by the net. Flume tank or wind tunnel studies are required to be certain that the flowmeter has been positioned in a representative part of the flow field. Tranter and Smith (1968) showed that for bridled circular nets, flowmeters should be positioned about one-third of the radial distance to the center. Flowmeter positioning in the mouth of unbridled circular nets is less critical than with bridled nets, since flow across the mouth is more uniform.

Each flowmeter should be calibrated both before and after a survey. This can be done either in a flume tank, or by towing it through a measured distance at different speeds. The number of revolutions is recorded separately for each speed to obtain a calibration constant (m/impeller revolution) for the towing speed being used. This factor will not be the same for each towing speed, even for the same flowmeter. Calibration tests should include speeds slow enough to define the friction point of each flowmeter being used.

Determining the volume fished during a plankton tow is far from simple, and can result in serious error if done improperly. A comparison of direct measurements of the volume being sampled by a plankton pump with estimates from an *in situ* flowmeter showed a 33% difference between methods (Brander and Thompson 1989). Comparisons of settled volumes of total plankton between the pumping system, a MOCNESS sampler, and a Longhurst–Hardy plankton recorder indicated errors of up to 68% in estimates of volume filtered, even though all three systems were equipped with calibrated flowmeters.

The CalVET (California Vertical Egg Tow) net (Figs. 4-8 and 4-9) provides an example of a net designed specifically for sampling anchovy or sardine eggs that are present in relatively high concentrations (Smith et al. 1985).

Figure 4-8. Retrieving a CalVET net during a vertical egg tow. Photo courtesy of K. Raymond, National Marine Fisheries Service.

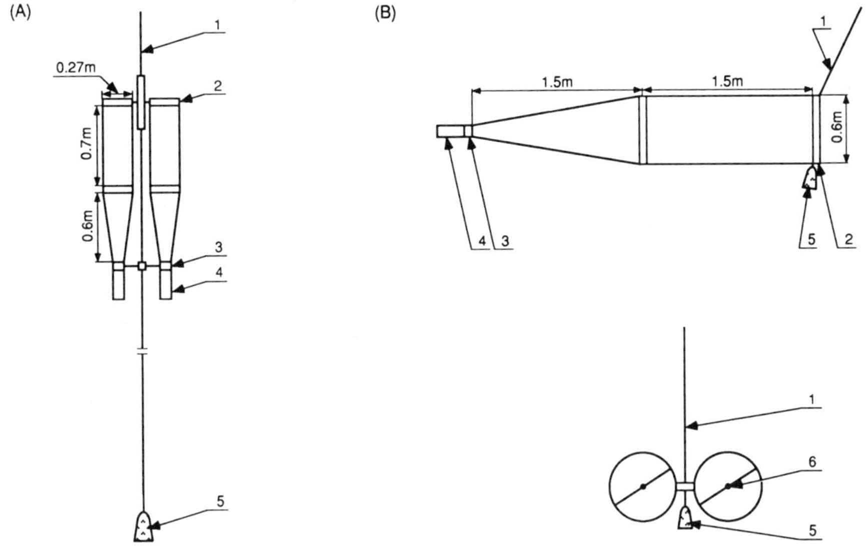

Figure 4-9. (A) Side view of a CalVET net during a vertical egg tow. (B) Side (upper figure) and front (lower figure) views of a bongo net during an oblique tow. (1) Hydrographic cable, (2) aluminum frame, (3) PVC coupler with rubber hose clamp, (4) nitex cod-end, (5) hydrographic weight, (6) flowmeter suspended in mouth of net.

The mouth opening is only 0.25 m in diameter, and is relatively small compared to the 1-m diameters frequently used in vertically deployed nets or the 0.6-m "bongo" nets used in double-oblique (Fig. 4-10) deployments. The smaller mouth opening reduces the sorting time required for each sample, allowing a greater number of stations to be occupied at a given level of cost or manpower availability. The mouth opening used will vary with the appli-

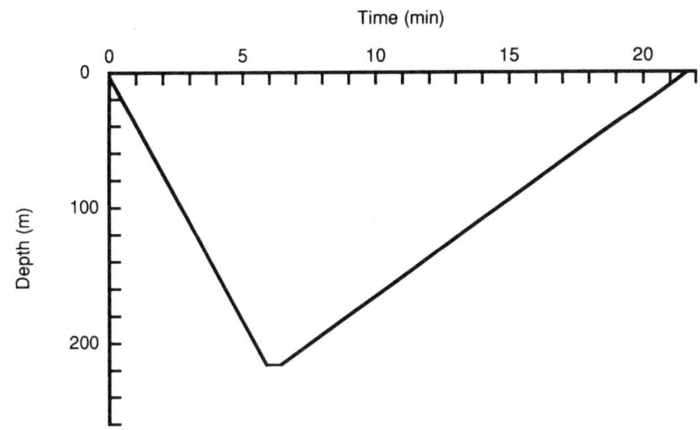

Figure 4-10. The ideal path for a double-oblique bongo net tow to 210 m.

cation, however, and would have to be increased when sampling a species whose eggs are sparsely distributed.

A 0.15-mm mesh size was selected for the net itself, since the diagonal aperature for this mesh is less than the minimum dimension of anchovy eggs (the target for these surveys). The net is deployed in a bongo-type (PAIROVET) frame, so that no bridles disturb the flow field over the mouth. The frame enables paired comparisons to be made for nets of different design and mesh size, although a second net is typically used as a stabilizing drogue even when such comparisons are not being made.

Since the goal of each tow is to obtain an estimate of mean egg density (no/m^3) throughout the entire water column (Equation 4.3), it is important that a constant volume of water be sampled at each depth. The vessel being used must be able to maneuver such that a 90° angle (± 15°) is maintained during the tow, and the winch employed must be able to retrieve the gear at a constant rate. It is particularly important that the net emerge from the water at right angles to the surface, since the eggs of many species are concentrated in the neuston layer.

Clogging can result in relative undersampling of shallow layers, and tows must be rejected where this is a problem. Clogging can be minimized by designing the nets with an elongated cylindrical section rather than a simple cone, providing a high ratio of open mesh area to mouth area. Tranter and Smith (1968) recommended combination cylindrical–conical nets with at least a 5:1 open mesh area to mouth area ratio to minimize losses in efficiency due to clogging, and the CalVET net has a ratio of 9.69:1 when 0.15-mm mesh is used.

Thorough washing of the net after each haul and replacement of nets that have built up residues in the meshes will help to minimize error caused by clogging. Deviations of flowmeter revolutions of 10% or more should be cause for exchange or careful washing of the net (Smith et al. 1985). Gehringer and Aron (1968) point out that without a second flowmeter, placed in the free stream outside the net, the investigator can only speculate on the possibility of clogging. Continuous monitoring of flowmeter data is a realistic alternative with current technology, however, and would also aid in detecting aberrant net behavior.

Ship drift, strong currents, and severe rolling by the vessel also result in nonuniform sampling of the water column, and tows should not be made when these conditions exist. When deployed in rough seas from a rolling vessel, the net can ascend in a series of rapid rises followed by sharp decelerations, potentially reversing directions and causing the contents of the net to be "pumped" out or extruded through the meshes . It is also possible that the net may reverse directions while sinking to depth, double-sampling some portions of the water column. As a result, the progress of the net through the water column should be monitored continuously with a depth sensor on the net. A bathykymograph (BKG) will provide a strip chart record of the tow, and time–depth recorders are available that are capable of temporarily storing this data and downloading it into a computer at convenient intervals. Real time monitoring of net depth is also possible using a depth sensor or conductivity–temperature–depth (CTD) recorder capable of transmitting data either acoustically or through a conducting wire. At the very least, the wire angle should be monitored at regular intervals, and tows repeated when they exceed a specified angle of stray.

Retrieval speed should be sufficiently rapid to eliminate stalling or reversals in the net trajectory as the vessel rolls, yet not cause eggs to be extruded through the meshes of the net. The CalVET net is lowered at a speed of 1–1.2 m/sec, and recovered at the same speed (Smith et al. 1985).

Double-oblique tows (Fig. 4-10) allow better control of the sampling path of the net than vertical tows, and are frequently used in conjunction with bongo nets in sampling eggs. The net is fishing when both sinking and being retrieved.

A bongo net is a pair of nets deployed in a single frame that is weighted such that no bridles are required (Fig. 4-9), and is capable of sampling larvae much more effectively than vertical sampling nets. Bongo nets may consequently be the preferred sampling gear when simultaneous sampling of both eggs and larvae is desired, although fewer stations can be sampled with a given amount of ship time and manpower.

The sampling path of a double-oblique tow is typically controlled by maintaining a constant winch speed and adjusting vessel speed so that the wire angle is maintained near a 45° angle at all times (Smith and Richardson 1977). This ensures uniform sampling of all depths. Smith and Richardson suggest lowering the net to the deepest sampling depth at a rate of 0.8 m/sec, and retrieving at a rate of 0.3 m/sec, recording the wire angle at 10 m (or 30 sec) intervals (Fig. 4-7).

Pump systems have also been developed for sampling eggs, and some are capable of sampling larger volumes of water per minute than many towed sampling gears (Solemdal and Ellertsen 1984). They are usually too slow for routine survey work, however, and are most useful for vertical profiling of egg distributions. Pump samplers are also of value for sampling near the bottom, or around submerged structures. A camera sampler, capable of allowing the survey scientist to identify and count eggs using photographs taken *in situ*, has also been developed (Houde et al. 1989), although its utility has yet to be fully evaluated over a range of survey conditions.

Sample Processing

Once aboard the vessel, the net is carefully rinsed to wash all material clinging to the net lining into a detachable plastic bucket (or nitex cod-end) at the end of the net. The contents of this bucket are then quickly placed in labled jars containing a fixative solution of 5% (2% formaldehyde) buffered formalin. The container should be filled to the top with the formalin solution, and the specimen fraction should take up no more than 25% of total container volume.

Sorting of the plankton samples usually takes place once ashore. Whenever possible, all eggs and larvae should be carefully removed from the plankton samples. Subsampling or "splitting" of the plankton samples can result in high variances when the eggs occupy a small portion of the sample, and subsampling should be limited to those instances where the samples contain large quantities of eggs. The Folsom plankton splitter is usually employed for subsampling (Smith and Richardson 1977).

When designing surveys and selecting the size of net to be used, it should be borne in mind that the sorting, identification, and staging of eggs can take considerable time. Sorting of bongo net samples can take about one day per sample (Smith and Richardson 1977), while about eight CalVET samples can be processed in the same period of time (Stauffer and Picquelle 1985). This is due to the small size of the CalVET sampler, and the relatively short duration of the tow.

The sorting of each plankton sample usually begins by draining off the preserving fluid to reduce the sorter's exposure to formaldehyde fumes. Total plankton volume is measured, and the sample is then placed in a beaker containing a dilute formalin solution. Small amounts of plankton are then spooned into petri dishes, and examined under a low-power (7–10×) binocular dissecting microscope. All eggs and larvae are removed using eyedroppers, pipettes, forceps, and wire loop probes. When sorting is complete, the remaining contents of the dish are poured into a beaker labeled "sorted," containing the original sample label together with fresh water and a few drops of concentrated formalin. This procedure is repeated until the entire sample (or aliquot if subsampling has been carried out) has been examined. Quality control can be accomplished by having a senior staff member sort through a predetermined fraction (e.g., 25%) of the contents of the "sorted" beaker, and resorting its entire contents if more than a single egg or larva is found. Alternatively, entire samples can be selected at random, and completely resorted. The combined results from sorting and quality control should be recorded on an Ichthyoplankton Sorting Form (Fig. 4-11).

To avoid significant reversal of the fixation process, the eggs, larvae and "sorted" sample should be placed in preservative as soon as possible. Eggs and larvae are usually transferred to a 3% formalin solution for permanent storage (Smith and Richardson 1977; Snyder 1983), although some investigators prefer to store larvae in 70% ethanol. Each group of eggs and larvae is placed in one or more vials and labeled. If experienced, the sorters may have already identified and separated some of the eggs and larvae by species, but many specimens require further examination by more highly trained ichthyoplankton taxonomists. The ancillary goals of the survey will determine whether all fish specimens or just the target species will be identified, and to what taxonomic level. Availability of manpower and financial resources, together with time deadlines imposed by the management process, will always

Page ____ of ____

ICHTHYOPLANKTON SORTING RECORD

Vessel/Cruise: ____________________ Gear/Mesh: ____________________

STATION	HAUL	NET	Sample DATE	No. Jars	Sorter	Eggs removed	No. Vials	Larvae Removed	No. Vials	QUALITY CONTROL Eggs	Larv.	Checker	Resort Y/N	Plankton vol.

EGG STAGING FORM

Cruise [ ][ ][ ][ ][ ] (1 2 3 4 5) Station [ ][ ][ ][ ][ ] (7 8 9 10 11) Gear [ ][ ] (12 13) Net [ ] (14) Haul [ ][ ][ ] (15 16 17)

Total Eggs [ ][ ][ ][ ][ ][ ] (21 22 23 24 25 26) Number Staged [ ][ ][ ][ ] (27 28 29 30)

Stage 1 ________ [ ][ ][ ] (32 33 34) Stage 7 ________ [ ][ ][ ] (36 37 38)

Stage 2 ________ [ ][ ][ ] (40 41 42) Stage 8 ________ [ ][ ][ ] (44 45 46)

Stage 3 ________ [ ][ ][ ] (48 49 50) Stage 9 ________ [ ][ ][ ] (52 53 54)

Stage 4 ________ [ ][ ][ ] (56 57 58) Stage 10 ________ [ ][ ][ ] (60 61 62)

Stage 5 ________ [ ][ ][ ] (64 65 66) Stage 11 ________ [ ][ ][ ] (68 69 70)

Stage 6 ________ [ ][ ][ ] (72 73 74) Stage 12 ________ [ ][ ][ ] (76 77 78)

Card [ ] (79) of [ ] (80)

Recorder ____________________ Date ____________________

Figure 4-11. Examples of forms used to record data obtained when sorting an ichthyoplankton sample and staging the eggs.

force hard decisions to be made and many specimens will usually be archived rather than being processed with the target species.

Egg Staging and Aging

Data on egg mortality rates and stage durations are required for estimates of daily egg production rate (Equation 4.1). To obtain this information, all eggs of the target species must be removed from the sampling vials and "staged" using carefully developed criteria (Table 4-3). Egg staging requires considerable time by experienced technicians, however, and it may be necessary to subsample the larger samples by selecting 100 eggs at random. Egg staging often involves different people than those doing the routine sorting, and the data obtained are recorded on an Egg Staging Form (Fig. 4-11).

The final goal is to assign an age (in hours or days) to each egg, and for optimum results the stages used to classify the eggs should each last less than 24 hours at ambient temperatures. This is critical for species such as the northern anchovy, where development is rapid and most eggs hatch within 3 days of being spawned at temperatures of 15–20°. Anchovy eggs are usually sorted into 11 stages of development (Fig. 4-3, Table 4-3), each lasting less than 12 hours at normal incubation temperatures (Lo 1985). In the case of walleye pollock in the Gulf of Alaska, however, eggs take 14 days to hatch at an ambient temperature of 5°C. Although many stages last a day or less, biomass estimates were obtained by combining several of these, and using six age groups of roughly 2 days each (Kim and Gunderson 1989).

Eggs in various states of disintegration are found in many samples, and with detailed examination, they can often be staged (Moser and Ahlstom 1985). Empty egg shells and eggs that have no identifiable morphological features are assigned to a "disintegrated " category (or categories if the range of possible stages can be narrowed). These can be assigned to a stage later, on the basis of the relative proportions of identifiable eggs in the sample.

Determination of stage durations and conversion of stage classification data into age frequencies require data from incubation experiments. Eggs can be obtained from a captive brood stock or from net hauls made at sea. The eggs can be incubated either ashore, or in shipboard laboratories, but facilities for accurate temperature control are required. Incubation experiments are usually run at several temperatures bracketing the range expected in the field, and the resulting data (Table 4-4 for example) fitted to an appropriate model. A separate model is often required for each stage, and is usually of the form

$$\text{Age} = e^{a - bT}$$

where T = temperature (°C)
a and b = constants for each developmental stage

Table 4-4. Average age (in hours) of anchovy eggs for each of 10 developmental stages (stage I not observed) at three temperatures, from laboratory experiments

	Temperature (°C)		
Stage	13.9	15.2	16.2
II	6.82	6.42	6.40
III	16.36	13.55	13.14
IV	24.91	21.12	20.18
V	35.88	30.7	25.56
VI	44.63	39.24	33.20
VII	49.64	47.13	38.56
VIII	52.5	52.07	43.64
IX	61.1	56.28	50.48
X	66.09	62.63	54.78
XI	69.79	65.65	56.10

Source: Lo (1985).

Alternative models may fit the data better, however. To obtain the predicted ages shown in Figure 4-4 for anchovy eggs, Lo (1985) used the equation

$$\text{Age} = 16.07\, e^{-(0.1145T + 0.098i)} i^{1.74}$$

where i = stage number and the same constants were used for all stages and temperatures.

Assigning an age to a given egg is simplified when spawning occurs within a restricted part of the day. Most spawning by northern anchovy occurs at night between 2100 and 0200 hours (Hunter and Macewicz 1980), making it possible to age their eggs with greater precision than for species that spawn evenly throughout the day.

Egg Mortality Rate (Z_{ij})

There are two basic approaches to estimating mortality rates, whether they pertain to adult fish or eggs. The first of these is the "catch curve" method, where it is assumed that egg production and mortality are constant over a given period of time, and the age frequency data for the eggs collected in a given stratum, survey, or series of surveys are aggregated to obtain a single "catch curve" (Fig. 4-12). In the second (cohort) method, a single cohort of eggs is identified on the basis of the age frequency data and followed over a sequence of cruises.

The catch curve method (which essentially creates a synthetic cohort) is most applicable for species such as anchovy, where the total incubation period is about 3 days (Picquelle and Stauffer 1985, Armstrong et al. 1988). The

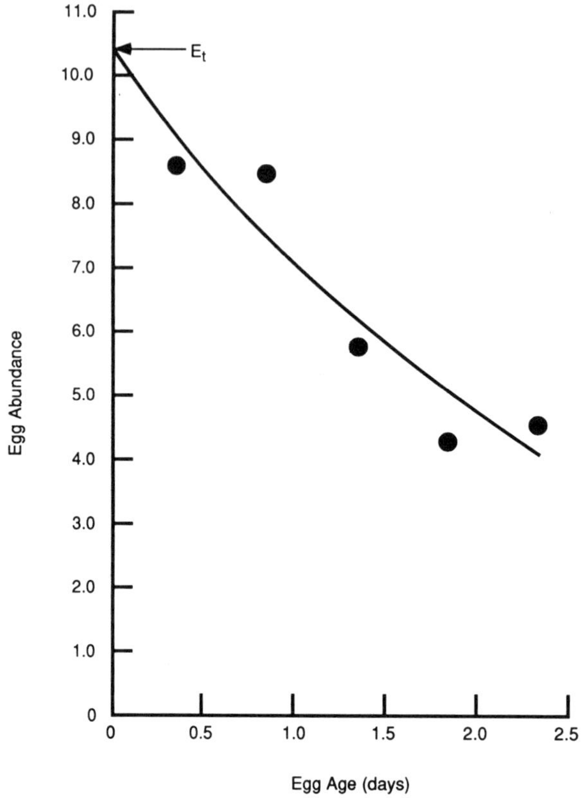

Figure 4-12. Fitting an exponential model (solid line) to data on egg abundance and age (days since fertilization). The intercept provides an estimate of daily egg production (E_t).

range of ages used to estimate Z (instantaneous daily egg mortality rate) is narrowed further by deleting ages that are too young to be completely available to the survey gear, or old enough that some individuals begin to hatch. This typically restricts the ages used to those between 5 and 60 hours old, and it is not unreasonable to assume that spawning intensity and mortality rate are constant over this period.

The coefficients of variation for estimates of Z for South African anchovies are typically quite high, however, ranging from 0.59 to 9.16 during 1984–1990 (Shelton et al. 1993), and could indicate a violation of the assumption that Z is constant. Recent studies of egg cannibalism in anchovy stocks off South Africa and Peru (Valdes et al. 1987; Smith et al. 1989) have provided evidence of density-dependent cannibalism rates, and such a mechanism could be expected to cause significant variations in egg mortality over relatively small scales of space and time. During a single survey in 1981, Z varied by over a factor of two for both sardine and anchovy in different geographic subdivisions of the Peruvian coast (Smith et al. 1989).

The catch curve method can easily be misused for boreal species such as walleye pollock, where the incubation period is about 14 days in Shelikof Strait. Spawning intensity will no doubt vary substantially over a 2-week period. Egg mortality rates are also dynamic, varying with predator abundance and the level of cannibalism. In Shelikof Strait, use of the cohort method showed that Z for 2- to 3-day-old pollock eggs varied systematically throughout the spring, declining from 0.43 in March to 0.091 in May (Kim and Gunderson 1989). In this situation, the catch curve should be restricted only to that range of age groups for which the assumption of constant mortality and spawning rate are valid. If cruises are spaced closely in time, an alternative approach would be to identify individual cohorts of eggs in the age frequency data, and follow the decline in their abundance during subsequent surveys.

The techniques used to estimate Z_{ij} from abundance at age data are similar, whether the data are obtained from a catch curve or by following a single cohort of eggs. Abundance at age (E_{ij}/d_{ij}) and age (t_{ij}) data are fitted to an exponential mortality model:

$$E_{ij}/d_{ij} = E_t e^{-Z_{ij}t_{ij}}$$

where all variables are defined as in Equation 4.1.

While Z_{ij} can vary between egg stages when using the cohort method, it is assumed to be constant for all stages (Z_j) in the catch curve method. Notice that estimates of Z_j and E_t are obtained simultaneously when using the catch curve method (Fig. 4-12).

A number of statistical techniques can be employed to fit the data for E_{ij}/d_{ij} and t_{ij} to this model. Where data are available for several age groups, weighted regression (linear or nonlinear) is commonly recommended (Picquelle and Stauffer 1985; Armstrong et al. 1988). The weights used in the regression should be based on the inverse of the variance around each estimate of abundance.

Population Estimation Models

Annual Egg Production Model

Three different methods (Table 4-1) have been developed to relate daily egg production (E_t) to the size of the adult population that produced them, and all require information on adult fecundity. The annual egg production (AEP) method (Equation 4.4, Table 4-1) is applicable only to determinate spawners, and requires an estimate of the total number of eggs produced per year by an adult female. This can be obtained by sampling the adults when vitellogenesis has progressed far enough that immature oocytes can be readily differentiated from those that will eventually be spawned, yet before any of the oocytes are hydrated. Once hydration has begun, it is difficult to determine how many batches of eggs have been released prior to collection of the sample, and some eggs may be extruded during the capture process. With the AEP method, total annual fecundity per female is related to total (integrated) annual egg production from the surveys (Equation 4.4, Table 4-1). All mature eggs in the ovary (potential annual fecundity) are counted when using this method. Atresia of oocytes prior to release can be corrected for by undertaking special collections and carrying out histological studies (Priede and Watson 1993).

When using the AEP method, multiple cruises are usually necessary to determine the shape of the spawning curve during each year surveyed. The shape of the spawning curve can then be estimated by fitting the egg production data by eye (Fig. 4-17), or by using a cubic spline function (Press et al. 1989).

Some authors have used a model-based approach, assuming, for example, that the spawning curve follows a unimodal sine curve (Picquelle and Megrey 1993), with a consistent shape and timing each year. A more simple model, based on the mean spawning curve for 1976–1983 was employed for the spawning stock of cod off Lofoten, Norway (Sundby and Solemdal 1984). The attraction of a model-based approach is that the AEP requirement for multiple cruises can be relaxed if the underlying model holds. The model could then be used to determine the shape of the spawning curve in any given year, and only a few surveys (only one in theory) would suffice to give estimates of total annual egg production.

Model-based approaches are to be avoided, however, unless there is a firm basis for assuming that the model being used will give reliable results. In the case of Lofoten cod, both the initiation and termination of the spawning period are remarkably stable (Fig. 4-2), and the median for the spawning curve had a standard deviation of only 2.2 days during 1968–1982 (Pedersen 1984). Nevertheless, significant error could be introduced by assuming that the mean spawning curve for 1976–1982 applies to each year in the series, and basing an AEP estimate of biomass on one or two cruises.

Daily Fecundity Reduction Model

The daily fecundity reduction (DFR) method (Lo et al. 1992) has also been used for determinate spawners, and requires sequential estimates of fecundity per adult female during the spawning season. A model is then fit to the data on the decline in population fecundity as spawning progresses, and used to relate the egg production on a given day to the corresponding drop in adult fecundity (Equation 4.5, Table 4-1). Population fecundity is estimated from the product of the proportion (G) of ovaries that are active (i.e., yolked oocytes are present) and the fecundity (F) of active females. All mature eggs are counted when using this method, and F includes both yolked and hydrated oocytes present in the ovary. Separate models are used to describe the decline in G and F as the spawning season progresses. In the case of Dover sole (Lo et al. 1993), linear models were sufficient to describe both declines:

$$G_t = c_0 + c_w w + c_t t$$

where w = female weight (g)
t = day of year
c_0, c_w, and c_t are constants, and

$$F_t = d_0 + d_w w + d_t t$$

where d_0, d_w, and d_t are constants.

These models were used to estimate the decline in fecundity (R_t) on day 50 (February 19), and relate it to a February 22–April 10 survey estimate of daily egg production (E_t).

The decline in fecundity (R_t) on day 50 was estimated from

$$R_t = \frac{d(\overline{F}_t\overline{G}_t)}{dt} = \overline{F}_t \frac{d\overline{G}_t}{dt} + \overline{G}_t \frac{d\overline{F}}{dt}$$

where all parameters are estimated for $t = 50$ and a female of average weight using the appropriate regression model.

Juvenile females are difficult to distinguish from adults in the resting state, so that both juvenile and adult females are included in the denominator of the fraction of females with active ovaries (G_t). It follows that the population estimate (Equation 4.5, Table 4-1) is an estimate of the total population vulnerable to the sampling trawl.

The DFR method was described only recently (Lo et al. 1992), and has consequently seen only limited application to date. Preliminary trials for Dover sole (Lo et al. 1993) indicate that the method shows considerable promise, and can provide reasonable levels of precision (CV = 0.19–0.38). The technique is most applicable to species which show a significant reduction in fecundity per female (R) over a relatively brief time interval, however. The fraction of Dover sole with active ovaries declined slowly (15% per month) during the spawning season, and adults had to be sampled over a relatively long expanse of time in order to detect a trend in R.

Daily Egg Production Model

The daily egg production (DEP) method can be used for either determinate or indeterminate spawners, and requires estimates of the "batch fecundity" (number of eggs that will be released by an average adult female during a single spawning event) and the fraction of females spawning per day (S). These are then related to the E_t value for that same spawning event to estimate population size (Equation 4.6, Table 4-1). Spawning events usually occur at 24 hour intervals in anchovies and sardines, and the method is consequently termed the "daily" egg production method. Time intervals greater or less than a day can be used, however, so long as the field-collected eggs from each spawning event can be identified by their ages. The fecundity estimate includes only those oocytes in the largest mode (batch) present in the ovary, or hydrated oocytes that are soon to be spawned when using this method.

Notice that only a single egg survey is required when using the DEP or DFR methods, offering a considerable reduction in survey costs over the AEP method. Fecundity data need only be collected a single time each year when using the DEP or AEP method, but a series of fecundity estimates is required when using the DFR method.

In the case of the DEP and DFR methods, all adult parameters (fecundity, fraction of females spawning, sex ratio) must be obtained by direct sampling of the adult population during the period the egg survey is being conducted. It is critical that the entire population be present in the survey area when the survey is conducted, or the population estimate will be biased. For this reason, egg surveys for both the DEP and DFR methods should be conducted near the peak of the spawning season. It must also be assumed that all adults are fully vulnerable to the sampling gear, and that survey catches are representative of the actual population.

Adult Fecundity

Whichever model is used to estimate population size, the general techniques used to collect ovaries and subsample the eggs from them are similar. The relationship between fecundity and length (or weight) is a close one, and large numbers of specimens are not required, so long they cover the entire length (or weight) range of the target population uniformly. The fecundity data are then fit to an appropriate linear or nonlinear model of the length–fecundity or weight–fecundity relationships, and the r^2 values obtained are generally high. Although the nature of the data will determine the model to be used, the relationships used most commonly are

$$F = aL^b$$

and

$$F = c + dw$$

where L = length (cm)
w = weight (g)
a, b, c, and d = constants

The linear model for the length–weight relation can be forced through the origin ($c = 0$), when fecundity/g is to be compared between geographic areas, years, etc.

Since hydration of the oocytes can significantly increase body weight, weights are usually taken after the gonad has been removed when using the batch fecundity method. All subsequent biomass estimates will consequently be made in terms of ovary-free wet weight. Hunter et al. (1985) determined that a sample of about 50 fish would provide CVs of less than 10% for the estimate of average batch fecundity for northern anchovy.

Fecundity estimates for individual specimens are usually obtained by subsampling the oocytes from the ovary on either a gravimetric or volumetric basis. The volumetric method involves placing the ovary in modified Gilson's solution to digest the ovarian tissue and disaggregate the oocytes. This solution can be obtained by mixing

100 ml	60% alcohol
880 ml	Water
15 ml	80% nitric acid
18 ml	Glacial acetic acid
20 ml	Formalin

The oocytes are separated from the ovarian tissue by washing them through successively finer meshed sieves. Once fully disaggregated, the oocytes are placed in a large beaker, and water added until about 2000 ml of oocytes and fluid has been obtained. The solution is then stirred using a magnetic stirrer and a subsample of about 5 ml obtained in a representative manner by drawing a pipette through the beaker. Care is required to sample the water column evenly with the pipette, and at least four replicate subsamples should be obtained from the same beaker to be certain that no bias occurs.

Some preliminary size frequency curves (Fig. 4-1) must be developed to delineate the size range of mature oocytes, and only oocytes that fall within this range are counted. The size frequency curves can be obtained by measuring individual oocytes with an ocular micrometer, or using image analysis equipment if it is available.

The number of oocytes in each subsample can be counted under a 10× binocular microscope to determine the number per milliliter. Alternatively, oocytes can be counted and measured using a digitizer linked to the microscope by a video camera system. If the CV for the subsample counts exceeds 10%, more subsamples should be obtained. With practice, all subsamples should provide estimates of oocytes per milliliter that agree within 10%. Total fecundity is obtained by extrapolating the mean number of oocytes per milliliter in the subsamples to the contents of the entire beaker.

Electronic devices (Witthames and Greer Walker 1987) have also been used to count and measure oocytes successfully, and all oocytes in the ovary can be counted. There is no need to subsample when this equipment is available.

Since Gilson's solution may destroy hydrated eggs, the gravimetric method must be employed when determining batch fecundity. This entails storing the ovaries from each specimen in 10% formalin upon collection. Ovaries showing signs of having already spawned part of a batch, indicated by the presence of hydrated oocytes in the lumen accompanied by significant numbers of hydrated eggs loose in the collection jar, should not be processed further.

Once back in the laboratory, the ovaries are weighed, and small samples of ovarian tissue (each containing 100–200 oocytes) removed from the ovary and weighed. Care must be taken to ensure that the number of oocytes per gram is relatively constant throughout the ovary, since serious bias can occur if this is not the case. The optimum number of tissue samples was two to three per ovary in the anchovy collected by Hunter et al. (1985), although more could be required if the number of eggs per gram varied with location in the ovary.

As with the volumetric method, a size range is established for those oocytes that are mature, and all oocytes that fall within this range are counted. This presents little problem when counting hydrated oocytes, since they can be distinguished by their large size and translucent appearance. Atretic hydrated oocytes from ovaries preserved in formalin are also distinguishable, and can be deleted from counts, so that atresia is not an important issue when estimating batch fecundity. The fecundity estimate for each specimen is determined by extrapolating the number of oocytes per gram from the tissue subsamples to the total ovarian weight.

Stereological methods, where the number of oocytes per unit ovarian volume is estimated by examination of histological slide preparations, have also been used to estimate fecundity (Emerson et al. 1990). While the use of these methods has been limited to date, they appear to offer distinct advantages in that

1. fecundity estimation, estimation of the incidence of atretic oocytes, and screening for postovulatory follicles can be carried out simultaneously, and
2. individual oocytes that are near the size threshold used to separate mature and immature oocytes can be examined carefully to determine their maturity status.

All fecundity values in Table 4-1 refer to an average value, and since fecundity varies with size, it is necessary to know the size–maturity schedule, and the size composition of adult females in the target population. With this information, the length (or weight)–fecundity relation can be used to obtain a weighted estimate of mean fecundity for all females, e.g.,

$$\overline{F} = \sum_i n_i m_i (aL_i^b) \Big/ \sum_i n_i m_i$$

where aL_i^b = predicted fecundity at length i

n_i = number of fish (or relative abundance) at length i

m_i = proportion mature at length i

Alternatively, data on m_i and n_i can be used directly to estimate the mean length or mean weight of a mature female. The length (or weight)-fecundity regression equation can then be used to predict the corresponding value of mean fecundity.

When using the AEP method, proportion mature and size composition data refer to the entire female population, and can be obtained from samples of the commercial catch, catch–age analyses, etc. In the case of the DEP and DFR methods, however, direct sampling of the adult population is required during the period the egg survey is being conducted. It is critical that the entire population be present in the survey area, or the estimates of mean fecundity will be biased.

Sex Ratio, Number to Biomass Conversion, Size and Age Composition

While all equations in Table 4-1 are given in terms of population size ($\hat{P}$ or $\hat{P}_s$), conversion to biomass ($\hat{B}$ or $\hat{B}_s$) is relatively straightforward. This can be done in a number of ways, the simplest of which is to multiply each population estimate by the appropriate mean weight. If a biomass estimate is the only goal, however, another common approach is to express both fecundity and sex ratio on a weight-normalized (fecundity per gram and percent females per mt, for example) basis. When this is done, all equations in Table 4-1 estimate biomass rather than population size.

Fecundity per gram can be estimated readily from the fecundity–weight relationship, given an appropriate mean individual weight estimate for the females in the population. This requires direct sampling of adults during the period when the egg surveys are being conducted in the case of the DEP and DFR methods, and it is critical that the midwater trawls and purse seines used are capable of providing unbiased estimates of both size composition and sex ratio. When this is the case, the direct sampling data can also be used to determine the age composition of the population, by subsampling the catches and obtaining age structures. The techniques used to process and subsample survey catches for species, size, and age composition are the same as outlined in Chapter 2, and all of the caveats in using trawl or purse seine catches to represent the actual population (Chapters 1 to 3) are also the same. Differential availability or gear avoidance behavior by size group, sex, or reproductive condition (immature, spent, resting etc.) will result in nonrepresentative catches.

When the AEP method is used, size composition, age composition, and sex ratio can be obtained by sampling commercial landings to obtain overall averages for the year. Estimation of sex ratio has proven to be surprisingly difficult in in the case of the DEP and DFR methods, however, since both methods require that representative samples of the adult population be obtained through direct sampling during the survey. This is particularly difficult during the spawning season, when fish are segregating and reaggregating on the basis of sex ratio and reproductive status. Northern anchovy, for example, exhibit a sharp decline in sex ratio toward the end of the diel spawning peak, with females making up only 25% of the fish captured during 2300–2359, as opposed to 43–52% during other hours of the day (Picquelle and Stauffer 1985).

The DFR method is subject to a further source of sampling bias, since it is difficult to distinguish immature (juvenile) females from those that are in the resting state. For this reason the r term in the denominator of Equation 4-5 is an estimate of the proportion of all females in the total population, and $\hat{P}$ is an estimate of the total population. Unless all juveniles are present in the survey area, and fully vulnerable to the sampling gear being used, however, the $\hat{P}$ estimate will apply only to a subset of the total population.

Spawning Fraction

An estimate of the proportion of adult females spawning per day (S) is required when using the DEP method (Equation 4.6), requiring histological examination of a representative sample of gonads. Gonads that are to be used in histological studies must be collected from several adult females, and stored in buffered 10% formalin (Hunter 1985) within an hour of capture. Picquelle (1985) examined the sampling requirements for estimating S, and found that it is generally more efficient to take samples from several trawl hauls, rather than sampling a large number of fish per haul. For northern anchovy she recommended sampling 15 ovaries per haul for at least 80 hauls to obtain a CV of 0.13 when S was 0.1. Sampling for adults was accomplished using a midwater trawl, targeting schools and aggregations that had been detected acoustically. Trawl samples were also made on the basis of alternative information on fish abundance, such as temperature and salinity data, or observable evidence of local concentrations (such as spawning products in plankton samples). Samples from each haul were used to estimate S, sex ratio and batch fecundity.

A different approach was used by Armstrong et al. (1988), with echo integration data being collected between all ichthyoplankton and trawl survey stations . The echo integration data were consequently available along continuous tracklines linking sampling stations (Figs. 3-25 and 4-15), and were used to weight the results from each station statistically. Trawl sampling was conducted with an Engels-308 midwater trawl with a mouth opening of about 15 m when towed at 3–4 knots. Trawl samples served a dual purpose, being used to identify the species composition of acoustic targets, as well as to estimate sex ratio, mean weight of adult females, and S for the DEP method. Fishing operations were carried out whenever acoustic targets were abundant. The anchovy dispersed into diffuse layers at night, and nighttime tows generally consisted of a double oblique haul through the scattering layer. A sample of 150 fish was obtained from each haul to estimate sex ratio and mean weight, and 25 of these were preserved for later determination of S.

Four oocyte maturity stages are associated with spawning; migratory nucleus stage, hydrated oocyte stage, early postovulatory follicle, and late postovulatory follicle. In principle any of these can be used to estimate the spawning fraction, so long as their duration is known, and representative samples can be obtained for the adults in that stage. Priede and Watson (1993) used the migratory nucleus stage for mackerel, but late-stage (day 1) follicles are typically used to estimate S for sardines and anchovies.

Careful histological analysis of the ovaries is necessary to obtain data on *S* from either the migratory nucleus stage, or from postovulatory follicles. Details of fixation, sectioning, mounting, and staining of the tissues are given in Hunter and Macewicz (1985b).

To use either the migratory nucleus stage, or postovulatory follicles to estimate *S*, it is necessary to divide the development, deterioration, and resorptive processes of the oocytes and follicles into a series of distinct histological stages, each with an assigned age. This requires a series of ovary samples taken at regular intervals, and can be accomplished either by spawning fish in the laboratory, or sampling spawning schools over a 24-hour period in the field. For optimal results however, the latter approach requires synchronous spawning with a brief, distinct spawning time, and is inherently less precise than spawning fish in the laboratory.

The histological features characterizing each stage may vary somewhat between species, but the general characteristics for anchovy (Hunter and Macewicz 1985b) will apply to most species:

Migratory nucleus stage: the nucleus is centrally located until just before hydration, when it migrates to the animal pole (Fig. 4-13A). Hydration begins about the time the nucleus has completed its migration and yolk globules have begun to fuse, forming yolk plates. The nucleus of a hydrated oocyte is not visible (Fig. 4-13B), since the nuclear membrane disintegrates after

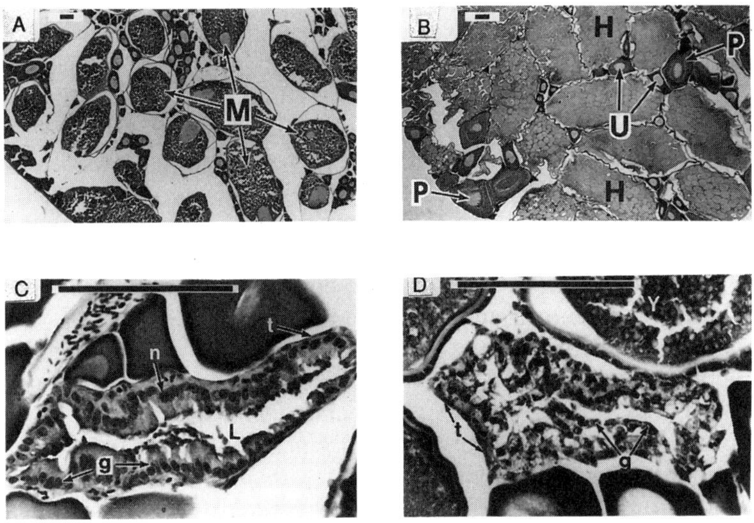

Figure 4-13. Development of anchovy ovaries at various magnifications. Bar = 0.1 mm. (A) Prespawning ovary showing oocytes in the migratory nucleus stage (M). (B) Imminent (<12 hours) spawning ovary with hydrated oocytes (H) still within their follicle layers (U = unyolked oocytes, P = partially yolked). (C) Early (day-0) postovulatory follicle with the nuclei (n) at the base of the granulosa cells (g). (L = lumen of follicle, t = theca). Elapsed time from spawning is 1–6 hours. (D) Day-1 postovulatory follicle. Elapsed time from spawning about 24 hours. From Hunter and Macewicz (1985b).

arriving at the animal pole, dispersing its contents into the cytoplasm. Migratory nuclei may be seen as early as 24 hours before ovulation, but hydrated oocytes in which all globules are fused to form plates do not occur earlier than 24 hours before spawning in anchovy.

Early postovulatory follicle: At ovulation, while the follicle is collapsing, the follicle cell layers appear cord-like, forming loose folds or loops. After spawning, the fully collapsed postovulatory follicle is a much more tightly folded or looped structure (Fig. 4-13C). It is irregular in shape, with a lumen which frequently contains eosinophilic granules. The granulosa cells in a new postovulatory follicle are characteristically columnar or cuboidal and in some cases have hypertrophied slightly. After spawning, the thecal layer is more clearly defined, adheres closely to the granulosa layer, and contains blood capillaries. There are no signs of follicle degeneration.

Late postovulatory follicle: The first sign of degeneration is the presence of a few pycnotic nuclei in the granulosa cells, followed shortly by the migration of lymphocytes into the follicle; finally a few vacuoles appear in the granulosa layer. By 19–28 hours after spawning, the regressing postovulatory follicle of an anchovy is greatly shrunken and has fewer folds (Fig. 4-13D), hence a less irregular form than a new postovulatory follicle. The lumen is much reduced, and may contain some granular material, although not as much as occurs in the lumen of younger follicles. The granulosa cells of the follicle no longer have the orderly alignment characteristic of younger follicles. Some of the nuclei of the granulosa cells are pycnotic, vacuoles are common, and only few of the cell walls are intact. The underlying layer of thecal cells is present, although less distinct than in early postovulatory follicles.

Classification of anchovy follicles becomes difficult beyond 28 hours as the follicles continue to degenerate, and by 48 hours they can easily be confused with the beta stage of atretic follicles. A postovulatory follicle 48 hours old is one-half to one-fourth the size of a 24-hour follicle, the lumen is much reduced or absent, and no eosinophilic granules are present. Cell walls are absent in the remaining granulosa layer tissue, and a few vacuoles or pycnotic nuclei may be seen. The theca is present, but is often indistinct as it becomes incorporated into the ovarian connective tissue stroma.

Females in the hydrated oocyte stage can be readily distinguished by macroscopic examination, but are are often difficult to sample representatively, evidently due to schooling behavior associated with spawning. A number of studies have demonstrated that hydrated female anchovies occur in disproportionately high numbers when sampled with purse seines and midwater trawls (Alheit 1985). Alheit suggests that hydrated females segregate from the rest of the population, either by depth or area, forming spawning schools that are dominated by males. These spawning schools are apparently more vulnerable to the sampling gear than the remainder of the population.

The oversampling problem often remains for early-stage postovulatory follicles. The "day-0" spawners shown in Figure 4-14 include northern anchovies with hydrated oocytes that will be spawned later that evening, as well

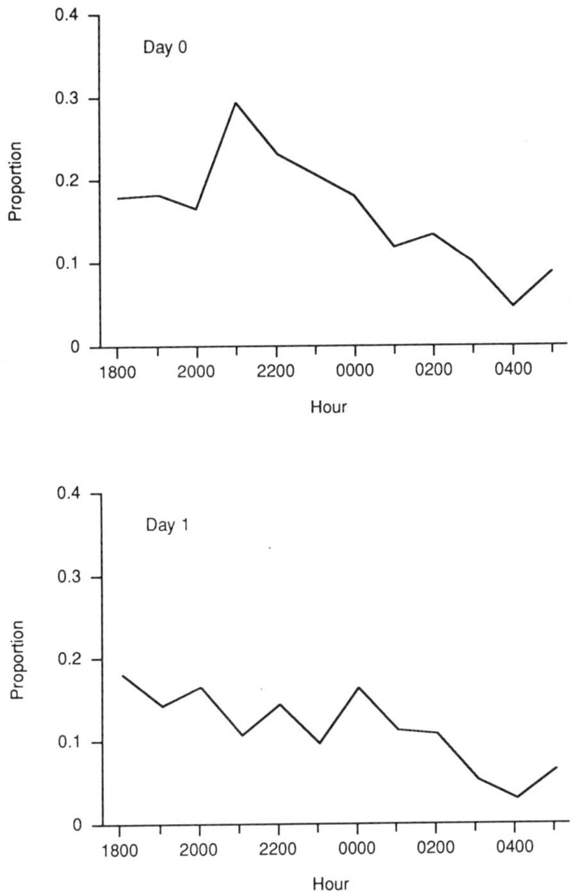

Figure 4-14. Proportion of day-0 (upper panel) and day-1 (lower panel) spawning anchovies, by hour, for midwater trawl surveys conducted during 1978–1980. After Picquelle and Stauffer (1985).

as those with postovulatory follicles that are 0–5 hours old (Picquelle and Stauffer 1985). The percent occurrence of day-0 spawners is consistently greater than that of day-1 spawners, and is highest during the hours of peak spawning (2100–2359). "Day-1" spawners, categorized as such on the basis the characteristics of postovulatory follicles that appear between 19 and 28 hours after spawning (Hunter and Macewicz 1985b), are used to estimate *S* for anchovies off California (Picquelle and Stauffer 1985) and South Africa (Armstrong et al. 1988). This is because the percentage occurrence for day-1 spawners is relatively stable between the hours of 1800 and 0300 (Fig. 4-14). The percentage occurrence of both day-0 and day-1 spawners falls off rapidly after 0300, as the error rate in classifying ovaries as to age increases. As a result, trawl sampling for adults is undertaken only during 1800–0300 (Hunter and

Macewicz 1985b). Oversampling of day-0 spawners is corrected for by replacing the observed number of day-0 spawners with the number that would have been observed (i.e., the number of day-1 spawners) had there been no oversampling. The corrected sample size then becomes

$$n = 2n_1 + n_2,$$

where n = total sample size
n_1 = number of day-1 spawners observed
n_2 = number of day 2+ spawners observed

DEMERSAL EGG SURVEYS

Most freshwater species and many marine species have demersal eggs. In some cases these eggs are adhesive and attached to either the substrate or to aquatic vegetation. Often, the eggs are deposited only on specific vegetation or characteristic spawning sites, presenting a much different sampling problem than surveys of pelagic eggs. In the case of Pacific herring, spawning sites are highly localized and patchy in their distribution, and random grids or surveys aimed at detecting them would waste considerable time and effort. However, herring spawning sites can be recognized by the clouds of milt, concentrations of birds, and egg deposition in the intertidal zone that are associated with spawning, and it is possible to locate each of them through regular aerial surveys. Furthermore, the spawning at any given site typically occurs over a matter of a few weeks, making it possible to enumerate the egg production for an entire year in a single census of each spawning site. As such, biomass estimates for this species represent a simplified case of the AEP method, where repeated surveys are unnecessary. The developing eggs are stationary until hatching (which requires 15 days in British Columbia), so that it is not necessary to estimate stage durations when conducting these surveys, and

$$E_a = \sum_{j=1}^{h} E_j e^{Z_j t_j}$$

where E_a = total annual egg production
E_j = total eggs observed in spawning site j
Z_j = instantaneous rate of daily egg mortality at site j
t_j = days elapsed since spawning occurred at site j
h = number of spawning sites

Population size is then estimated from Equation 4.4 (Table 4-1).

While some aspects of population estimation are simplified when conducting demeral egg surveys, estimation of E_j is far more complicated than is the case with pelagic surveys. It is relatively straightforward to delineate the bounds of each spawning site, and determine their length and width using grappling hooks, viewing boxes, snorkeling, and SCUBA. Estimating egg density presents significant problems, however.

A systematic series of transects can be set up across each spawning site, and quadrats (typically 0.1 to 0.5 m^2) placed at fixed intervals along each transect by SCUBA divers. Schweigert et al. (1990) used optimal sampling theory to determine that a sampling intensity of 3 transects/km with a minimum of 1 quadrat sample per 25 m of transect length would result in estimates of mean egg density with a CV no greater than 25% for herring surveys off British Columbia.

Enumeration of the eggs within each quadrat is difficult, since the eggs are attached to vegetation and substrate. Direct estimation is accomplished by removing all vegetation and substrate covered with eggs from each quadrat, immersing the samples in Gilson's fluid, and returning them to the laboratory. The Gilson's solution is then drained off, and each sample placed in a potassium hydroxide solution to dissolve the kelp and loosen the eggs from any remaining vegetation or substrate. Repeated washing and seiving are required to separate and clean the eggs, which can then be enumerated using either volumetric or gravimetric subsampling techniques. The costs and manpower associated with doing this are prohibitive in areas where herring are abundant, however, and considerable effort has gone into the development of visual techniques for estimating egg density (Schweigert et al. 1990).

Visual Models of Egg Density

Visual models are based on diver observations of parameters such as visual estimates of egg density, number of egg layers covering vegetation, vegetation type, total wet weight of vegetation, percent coverage by vegetation etc. Empirical models are then developed, using data from a subsample of quadrats where the number of eggs present has been estimated through laboratory processing and counting.

Biggs and Funk (1988) describe the development of such an empirical model, beginning with a fully-parameterized multiplicative model:

$$Y_{ijk} = e^{(\alpha + D_j + V_k)} X_{ijk}^{B_{jk}} e^{\varepsilon}$$

where Y_{ijk} = ith laboratory egg count for diver j and vegetation type k, based on a subsample of 2.5% of total quadrats

α = constant

D_j = effect of diver j

V_k = effect of vegetation type k

B_{jk} = parameter controlling functional relation between visual diver estimate X_{ijk} and the laboratory egg count

X_{ijk} = ith visual estimate for diver j and vegetation type k

ε = normally distributed random variable with mean 0 and variance σ^2

Stepwise multiple regression was then used to determine subsets of the main effects, which would explain the maximum amount of variability with the smallest number of parameters, and it was found that the model

$$Y_{ik} = e^{\alpha} X_{ik}^{B_k} e^{s^2/2} \tag{4.7}$$

where s^2 is the mean square residual error, explained 86% of the variance in the 1988 data from Prince William Sound, Alaska.

Rather than extrapolate the results beyond the data collected, a separate model is developed each year. Schweigert et al. (1990) found that extrapolation of a prediction model that accounted for 84% of the total variation in observed egg numbers to data from areas and years that were not used in model development yielded predictions that ranged from a 9% overestimate to a 41% underestimate.

VARIANCE ESTIMATES

Variance of Adult Parameters

Adult parameters include fecundity (F), sex ratio (r), mean weight (W), fraction spawning per day (S), and fraction of females with active ovaries (G). The exact combination of these required depends on whether the AEP, DFR, or DEP method is being used (Table 4-1).

In the case of the AEP method, both fecundity and sex ratio data are collected independent of the egg survey, often from samples of the commercial catch. The variance of the fecundity estimate can be determined from the length–fecundity or weight–fecundity regression parameters, and Var(r_s) from the variance of a binomially distributed variable. Nevertheless, difficulties in estimating Var($\int E_t dt$) usually preclude variance estimation in the case of the AEP method. If the shape of the annual production curve is determined in an objective way, however (e.g., connecting the E_t values with straight lines or using a cubic spline function), the bootstrap method could be used to estimate Var($\int E_t dt$).

In the case of the DFR and DEP methods, adult parameters are obtained from trawl hauls made with the specific goal of estimating them, and the results for each haul are weighted by factors such as catch size, sample size or acoustic estimates of abundance in areas adjacent to the haul. Since the adult parameters often vary between strata, it is usually necessary to obtain weighted means for each stratum. For a given trawl parameter μ_{ij} (where $\mu_{ij} = r_s$, S, or W in stratum i and transect j) Armstrong et al.(1988) estimated $\bar{\mu}_i$ from

$$\bar{\mu}_i = \frac{\sum_j w_{ij} \mu_{ij}}{\sum_j w_{ij}} \tag{4.8}$$

where w_{ij} = weighting factor = $L_{ij}D_{ij}$
L_{ij} = length of transect j
D_{ij} = mean density of adults (g/m^2 for r_s, numbers/m^2 for S and W), based on acoustic observations on transect j.

and its variance from

$$\mathrm{Var}(\overline{\mu}_i) = \frac{(T_i/T_i - 1)\sum_j w_{ij}^2(\mu_{ij} - \overline{\mu}_i)^2}{\left(\sum_j w_{ij}\right)^2} \tag{4.9}$$

where T_i = number of transects in stratum i.

Batch fecundity can be estimated in a number of ways, either using the fecundity–weight regression to estimate $\overline{F}_b$ directly from the mean weight estimate (Armstrong et al. 1988) or using the regression equation to estimate F_b for each mature female subsampled (Picquelle and Stauffer 1985). In either case, the variance of this estimate must take the component of variation arising from estimation error in the fecundity–weight regression into consideration. Armstrong et al. (1988) define the variable $\overline{Q}$ as the ratio $\overline{W/F}$, and give its normalized variance as

$$\frac{\mathrm{Var}(\overline{Q})}{(\overline{Q})^2} = \frac{1}{(\overline{Q})^2}\left(\frac{1 - b\,\overline{Q}}{\overline{F}}\right)^2 \mathrm{Var}(\overline{W}) + \frac{1}{(\overline{F})^2}\left[\frac{S_{FW}^2}{n_{FW}} + S_b^2(\overline{W} - \overline{W}_r)^2\right]$$

where b = slope of fecundity–weight regression
S_{FW}^2 = residual sums of squares of the regression
S_b^2 = variance of b
n_{FW} = number of observations in the regression
$\overline{W}$ = estimated mean female weight in the population
$\overline{W}_r$ = mean female weight in the regression sample

Variance of E_t

The estimation of daily egg production (E_t) depends on whether the catch curve or cohort approach is used to estimate Z_{ij}, as well as the manner in which the data are weighted in the regression model. As a result, estimation of the variance of E_t varies between applications. When using the catch curve method, the variance estimates for the regression parameters from the model: $E_{ij}/d_{ij} = E_t e^{-Z_{ij}t_{ij}}$ can be used to estimate E_t and its variance directly (Picquelle and Stauffer 1985).

Armstrong et al. (1988) took a different approach, where values of egg abundance for the different batches at each station were divided by $e^{-Z_{ij}t_{ij}}$ to provide up to three estimates of the daily egg production. These were then averaged to give a station value for E_t. Stratum mean values for E_t were obtained by averaging all station values for a given transect (Fig. 4-15), and then obtaining a weighted mean ($\overline{E}_t$) with weights equal to transect length (as in Equation 4.8). The normalized variance of this estimate was determined using equations for the variance of a weighted mean, with an additional term for the error associated with the estimation of egg mortality:

$$\frac{\mathrm{Var}'(\overline{E}_t)}{\overline{E}_t^2} = \frac{\mathrm{Var}(\overline{E}_t)}{\overline{E}_t^2} + (\bar{t}^2)\,\mathrm{Var}(Z)$$

where $\mathrm{Var}'(\overline{E}_t)$ = adjusted variance of $\overline{E}_t$
$\mathrm{Var}(\overline{E}_t)$ = unadjusted variance of $\overline{E}_t$ obtained from an equation similar to Equation 4.9
$\bar{t}$ = weighted mean age of eggs over the survey area

$$\mathrm{Var}(Z) = \frac{S_{yt}^2}{\left(N\Big/\sum_t w_t\right)\sum_t w_t(t - \bar{t})^2}$$

where S_{yt}^2 = variance around regression line for $\ln(E_{ij})$ versus t
w_t = weighting factor for $\ln(E_{ij})$ data = $1/\mathrm{Var}[\ln(E_{ij})]$
t = mean age in each age class
N = number of age classes in the regression

Lo et al. (1993) used the bootstrap method to estimate $\mathrm{Var}(E_t)$ for Dover sole. In their application, a data set of 116 plankton hauls was resampled,

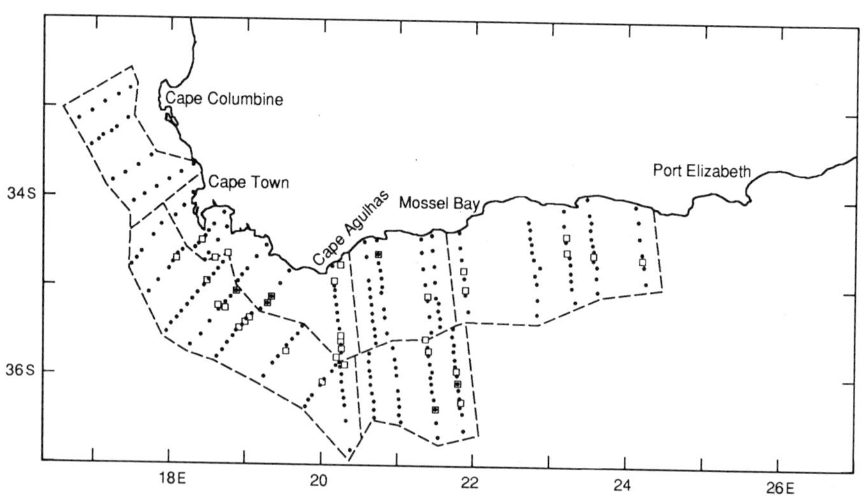

Figure 4-15. Location of egg stations (dots) and midwater trawl catches (squares) off South Africa in 1985. Stratum boundaries are indicated by dashed lines. After Armstrong et al. (1988).

with each iteration including a random sample (with replacement) of 116 records from the original data set. Each record contained data on the number of eggs observed in a given haul, by age group. Daily egg production (E_t) was then estimated for each iteration, using the methods routinely used for the estimation of development time and mortality rate. The point estimate and variance for E_t were determined from the mean and variance of 500 iterations of this process.

Variance of $\hat{B}$

By defining $Q = W/F$, Armstrong et al. (1988) expressed the biomass estimate for the DEP method as

$$\hat{B} = \frac{\overline{E}_t \overline{Q}}{\overline{S}\bar{r}_s}$$

Since the adult parameters were all determined from the same sampling program, the estimates were not independent, and the normalized variance of $\hat{B}$ was estimated using the delta method (Seber 1973):

$$\begin{aligned}\frac{\text{Var}(\hat{B})}{(\hat{B}^2)} &= \frac{\text{Var}(\overline{E}_t)}{(\overline{E}_t)^2} + \frac{\text{Var}(\overline{Q})}{(\overline{Q})^2} + \frac{\text{Var}(\overline{S})}{(\overline{S})^2} + \frac{\text{Var}(\bar{r}_s)}{(\bar{r}_s)^2} \\ &\quad + 2\left[\frac{\text{Cov}(\overline{E}_t\overline{Q})}{(\overline{E}_t\overline{Q})} + \frac{\text{Cov}(\bar{r}_s\overline{S})}{\bar{r}_s\overline{S}} - \frac{\text{Cov}(\overline{E}_t\bar{r}_s)}{\overline{E}_t\bar{r}_s}\right. \\ &\quad \left. - \frac{\text{Cov}(\overline{E}_t\overline{S})}{\overline{E}_t\overline{S}} - \frac{\text{Cov}(Qr_s)}{Qr_s} - \frac{\text{Cov}\ \overline{Q}\,\overline{S}}{\overline{Q}\,\overline{S}}\right]\end{aligned} \tag{4.10}$$

Since all parameter estimates were transect-based, covariances for any pair of parameters in stratum i were estimated from:

$$\text{Cov}[\bar{\mu}(1)_i, \bar{\mu}(2)_i] = \frac{(n_i/n_{i-1}) \sum_j w(1)_{ij} w(2)_{ij} [\bar{\mu}(1)_{ij} - \bar{\mu}(1)_i][\bar{\mu}(2)_{ij} - \bar{\mu}(2)_i]}{\sum_j w(1)_{ij} \sum_j w(2)_{ij}}$$

The bootstrap method offers a simpler approach to variance estimation. Lo et al. (1993) applied this technique in estimating the variance of adult parameter estimates for Dover sole when using the DFR method. Two data files were extracted from the adult fish data sets, to be bootstrapped for estimates of the variances of R, W, and r. Data required to compute r, W, and G were contained in one data file, while the data required to compute weight-specific fecundity ($=F/W$) were contained in a separate data file. Both data files contained identification of the haul, (i.e., year and haul number) for cross-reference. In the first data file, each record contained sample data: total female weight sampled, number of females, total sample weight, number of fish in the sample, number of fish with active ovaries, and number of females with inactive ovaries. In the second data file, each record contained fecundity information on individual fish.

For each stratum, a fixed number of tows (corresponding to the number of tows actually made in that stratum during the survey) was selected at random (with replacement) from the first data file. Any one selected tow was further checked to see if any fish had been sampled for fecundity. If so, fish were selected with replacement until the sample size corresponded to the original number of fish sampled for fecundity. If not, the resampling process was continued until this sample size had been obtained. The total number of fish bootstrapped for each stratum corresponded to the actual number of fish sampled there. For each iteration, r, W, $\overline{F}_t$, and R_t (Equation 4.5) were computed using standard estimation methods. A total of 500 iterations was made, and the mean and variance of each parameter were estimated from the results.

The bootstrap method has also been used to estimate the mean and variance of egg density estimates from demersal egg surveys for Pacific herring. Schweigert (1993) generated 400 bootstrap samples of a two-stage sampling process. For each spawning site, transects were selected at random (with replacement) and, within transects, individual quadrat samples were selected at random. For each quadrat selected, the estimates of egg density observed on the vegetation and bottom substrate were summed to obtain an estimate of total egg density. However, no attempt was made to bootstrap the egg prediction models or estimates of eggs on bottom substrate.

EXAMPLES OF SUCCESSFUL APPLICATIONS

Sardines and Anchovies

Sardines and anchovies always comprise a major portion of the world's fish catch, and most of them are caught in current systems that produce upwelling. Reliable assessment of these resources has often proven difficult, however, owing to difficulties in age determination. Acoustic surveys have been successfully carried out for these species, but early efforts were plagued by uncertainties in calibration, target strength estimation, and the vessel avoidance problems associated with acoustic surveys of near-surface schools. Both sardines and anchovies are indeterminate spawners, making application of the traditional annual egg production method impossible.

A series of ichthyoplankton surveys was initiated in the California Current region in 1949 under the CalCOFI (California Cooperative Oceanic Fisheries Investigations) program, but provided only a relative index of larval abundance. In 1976, when the United States assumed jurisdiction over the resources within its 200 mile limit, no reliable methodology was available to provide estimates of absolute biomass for the anchovy stocks within this zone.

New methods of stock assessment were clearly needed, and the daily egg production method was developed in response. Its first successful application was in the California Current system in the early 1980s. Anchovy eggs were collected using a CalVET net at stations spaced at 4 nm (7.4 km) intervals along onshore–offshore transects that were 10–20 nm (18.5–37.1 km) apart (Fig. 4-6A).

Adult parameters were determined through direct midwater trawling (Fig. 4-6B). Three strata were defined, due to latitudinal variability in egg density, egg mortality rates, and adult (percent spawning per day, mean weight, and batch fecundity) parameters.

The results of the early surveys (Table 4-5) served to establish the feasibility of using the DEP approach to estimate biomass, and showed that the variance of the resulting biomass estimate was within acceptable limits. Approximate (2 CV) 95% confidence limits ranged from ±29–51% of the spawning biomass estimate during 1980–1983.

The DEP technique quickly replaced the larval surveys formerly used to monitor anchovy stocks in the California Current area, and was soon adapted to sardine and anchovy stocks in upwelling areas such as the Humboldt Current off Peru, the Benguela Current off South Africa, and the Canary Current off Spain and Portugal. In addition to providing biomass estimates for the stocks in these regions, the data obtained when employing the DEP method have provided considerable insight into interannual variability in the reproductive biology (spawning frequency, batch fecundity etc.) and egg mortality rates (Z) of the fish in these systems (Alheit 1993). This has led in turn to a

Table 4-5. Egg production parameter estimates for California Current anchovy, 1980–1983

	1980		1981 (February)		1981 (April)		1982		1983	
	Est.	CV[a]	Est.	CV	Est.	CV	Est.	CV	Est.	CV
Daily egg production ($\times 10^{12}$ eggs/day)	26.34	0.111	20.96	0.101	12.59	0.087	13.51	0.237	17.25	
Fraction of mature females spawning/day	0.142	0.125	0.106	0.122	0.125	0.092	0.120	0.038	0.094	
Sex ratio by weight (females/total)	0.478	0.120	0.501	0.063	0.495	0.051	0.472	0.081	0.549	
Batch fecundity (eggs/batch)	7,751	0.075	8,329	0.052	8,846	0.045	10,845	0.047	5,297	
Average female weight ($\times 10^{-6}$ t)	17.44	0.055	13.37	0.039	16.20	0.029	18.83	0.019	11.20	
Spawning biomass (10^3 t)	870	0.225	635	0.183	372	0.147	415	0.257	652	0.211

[a]CV, coefficient of variation.

Source: Hewitt (1985).

better understanding of the processes that determine year class strength in upwelling areas.

Daily egg production and acoustic surveys have been employed simultaneously in the Benguela Current area (Armstrong et al. 1988, Jolly and Hampton 1990). Acoustic data were collected along transects that had been selected quasirandomly (Fig. 3-25). The vessel was stopped at 9.3 km intervals (increased to 18.6 km in areas of zero egg abundance), and vertical egg sampling conducted using a CalVET net (Fig. 4-15). Midwater trawl sampling for adults served to provide both target identification data for the acoustic surveys and adult spawning parameters for the DEP method. Length frequency data and otoliths were collected from each trawl catch, in order to provide estimates of the size and age composition of the spawning stock. As described earlier in this chapter, the acoustic data also provided relative abundance data for adults that were used to weight the adult spawning parameters obtained from each trawl sample.

The results of the surveys of the Benguela anchovy (*Engraulis capensis*) stock enabled the investigators to obtain two separate estimates of the biomass from a single survey (Table 4-6). The precision of the DEP method was much lower than for the acoustic estimates, largely due to imprecision in the estimation of egg mortality rate (reflected in the high CVs for the daily egg

Table 4-6. Egg production parameter estimates for Benguela Current anchovy, and corresponding acoustic estimates of biomass

	1985		1986	
	Est.	CV[a]	Est.	CV
Daily egg production (eggs/day/m^2)	326.5	0.20	487.6	0.25
Average female weight (W,g)	14.20		11.70	
Batch fecundity (*F*, eggs/batch)	7,991		5,627	
Q (= W/F)	—	0.05	—	0.06
Sex ratio by weight (females/total)	0.481	0.06	0.553	0.02
Fraction of females spawning/day	0.186	0.31	0.095	0.21
2Σ covariances	—	0.14	—	0.09
Area (km^2)	106,708		109,841	
Egg production biomass estimate (10^3 t)	614	0.41	2,006	0.35
Biomass from acoustics (10^3 t)	975	0.16	1,747	0.14

[a]CV, coefficient of variation.

Source: Armstrong et al. (1988).

production estimates) and the fraction of females spawning per day. Approximate 95% confidence limits for the DEP biomass estimate were ±70–82%, much higher than those attained in the California Current region. Armstrong et al. (1988) attributed this to the relatively low (about 300) number of CalVET stations occupied, but estimated that overall survey effort would have to be at least doubled in order to reduce the CV for the spawning biomass estimate from 35 to 20%.

There was no evidence that either the DEP or acoustic method was biased, since the point estimates for the two methods agreed relatively well. The target strength value used in the acoustic estimate was based on a target strength–length relation reported for herring, however, and the acoustic estimates must be regarded as having a potentially large and unknown bias until direct estimates of anchovy target strength are made.

The results from direct surveys of southern Benguela anchovy biomass initially allowed the identification of appropriate levels of yield for the resource, and catches were increased from 300,000 tons in 1986 to 600,000 tons in 1987, largely on the basis of survey results (Armstrong et al. 1988). Biomass estimates from the egg production method, acoustics, and catch-at-age data have been integrated into the modeling and management of the Benguela anchovy stock, and have played an important role in setting harvest levels since 1987 (Shelton et al. 1993).

Atlantic Mackerel

During the mid 1960s there was a rapid increase in catches of mackerel (*Scomber scombrus*) in European waters, due to advances in fishing technology, a crash in herring stocks, and a switch in fishing effort. By 1979, mackerel accounted for 42% of the total UK fin-fish catch (Priede and Watson 1993).

Two major stocks of mackerel have been distinguished in the Northeast Atlantic. The Western stock spawns west of the British Isles (predominantly along the edge of the continental shelf) during April–June, while the North Sea stock spawns in the northern North Sea in May and June. Both stocks migrate extensively over the European continental shelf (including the North Sea and English Channel), and mix during the nonspawning season. The North Sea stock was severely depleted as of 1990, whereas the western stock had a potential yield of about 500,000 tons (Priede and Watson 1993).

Since Atlantic mackerel lack a swimbladder, they are poor candidates for acoustic surveys, and tagging experiments were carried out in 1969–1976 in order to estimate the size of both the North Sea and Western stocks. These failed to provide reliable biomass estimates for the Western stock, however (Lockwood 1988), and in 1977 a series of surveys were carried out in order to estimate the biomass using the annual egg production (AEP) method.

Since the AEP method requires multiple surveys during the spawning season, the combined resources of five nations could provide estimates of biomass only at 3-year intervals. In addition, the assumption that mackerel are determinate spawners (required for the AEP method) was never substantiated, and the validity of the results was open to question. As a result, the decision was made to apply both the AEP and daily egg production (DEP) methods in 1989.

Five egg surveys were carried out using double-oblique tows with a Gulf III high-speed plankton sampler to provide the information required for the AEP method (Priede and Watson 1993). Plankton stations were spaced at 0.5° latitude and longitude giving a 30 by 18 nautical mile grid (Fig. 4-16). The survey area was adjusted to follow changes in the area of egg production over the course of the spawning season. The plankton survey for the peak of the spawning period also served to provide the egg data required for the DEP method (Fig. 4-17).

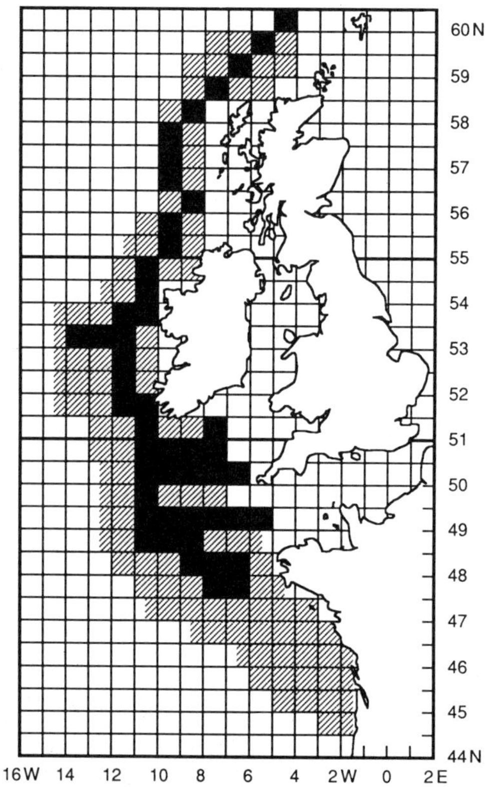

Figure 4-16. Location of egg stations (shaded squares) and midwater trawl catches (black squares) in the northeast Atlantic during May–June 1989. Two egg stations were sampled in each of the shaded grid squares, with some additional replicates for variance estimation. The half-shaded squares were sampled only once in the half indicated. Three strata were delineated, as indicated by the darkened horizontal lines. After Priede and Watson (1993).

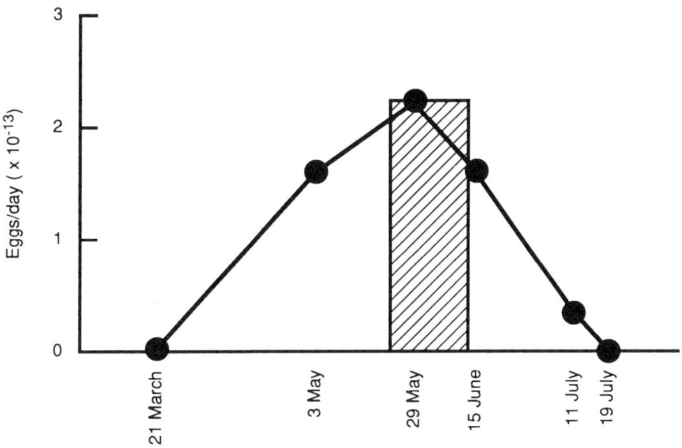

Figure 4-17. Eggs produced per day (E_t) by the Western stock of mackerel, during six surveys during 1989. The solid line indicates the estimated shape of the annual spawning curve, which was integrated to apply the AEP method of biomass estimation. The rectangle indicates the cruise during which data on adults were collected for application of the DEP method. After Priede and Watson (1993).

Estimates of eggs produced per day (E_t) were based only on stage 1 eggs, and egg mortality (Z in Equation 4.1) was ignored. The duration for this stage varies between 27 hours at 16°C and 68 hours at 9°C, and failure to account for mortality results in some underestimation of biomass.

An additional vessel employed a midwater trawl with an 80 by 100 m opening to sample adult mackerel and obtain data on the fraction spawning per day and batch fecundity. Midwater trawl hauls were 1 hour in duration, the net being placed near the bottom for the first half hour and near the surface for the second half, thus sampling the entire water column. Latitudinal trends in spawning fraction and batch fecundity were evident, and the survey area was divided into three strata as a result (Fig. 4-16). Length measurements and otoliths were also collected from the midwater trawl catches, in order to determine the size and age composition of the stock.

Potential annual fecundity was determined from a sample of 100 fish captured early in the spawning season, before they had released any eggs. The size threshold above which oocytes should be counted was difficult to establish, since there was often no clear break in size distribution between immature oocytes and those that would be spawned that year. It was recognized that mackerel might be indeterminate spawners, and that some of the immature oocytes could later undergo oogenesis, so that the size threshold finally selected was somewhat arbitrary. For the 1989 fecundity estimates, oocytes larger than 130 mm were counted as contributing to the potential annual fecundity. This diameter is much smaller than the start of true vitellogenesis, a stage reached at about 300 mm. The threshold adopted consequently allows for some degree of development of resting oocytes during the course of the season.

Batch fecundity was determined by counting the number of hydrated oocytes in females obtained in midwater trawl hauls. A total of 30 ovaries per haul was collected for this purpose, and stored in 10% formalin. Once back in the laboratory, the ovaries were screened to see which ones contained hydrated oocytes and were suitable for determination of batch fecundity.

Histological slides of ovaries from 1330 fish were analyzed for the incidence of the four main oocyte maturity stages associated with spawning. Immature 1- and 2-year-old fish with nondeveloping ovaries were excluded from the samples collected for this purpose. The fraction of mature females spawning per day was based on the incidence of fish in the migratory nucleus stage, tentatively assuming that this stage had a duration of 24 hours.

Atresia was evaluated using the same histological sections that had been prepared to determine the spawning fraction. The results indicated that mackerel abort one or more batches of oocytes through atresia during the spawning season, representing a loss of about 5% of potential annual fecundity. No adjustments for atresia were made to the estimates of potential annual fecundity when using the AEP method, however.

The sex ratio (proportion of females) in midwater trawl hauls varied between 0.53 and 0.54 within the three latitudinal strata. The average fraction of females in the commercial catch was 0.49 during 1982–1989, however, and a value of 0.5 was adopted for biomass estimates.

The resulting estimates of prespawning adult biomass in 1989 were 2.19 million tons for the DEP method and 1.87 million tons for the AEP method. This relatively close agreement suggested that the estimate of potential annual fecundity was approximately equal to actual annual fecundity, and that any maturation of oocytes classified as immature was more than compensated for through atresia.

A variance estimate for the DEP estimate of biomass was calculated using variance estimates for daily egg production, batch fecundity, and percent spawning per day, and gave a CV of 0.26 (approximate 95% confidence interval = biomass estimate ±52%). No variance estimate was obtained for the AEP method. The CV for total annual egg production (E_a) was estimated from 110 rectangles in which replicate samples were taken, and was found to be 1.3. The most important source of variance came from the estimate of daily egg production (E_t), which had a CV of 0.18 for the survey made at the peak of the spawning season. Trials showed that the number of stations sampled per vessel day could be roughly doubled by employing a CalVET vertical tow net instead of the Gulf III sampler, with substantial reductions in variance.

While changing from the AEP to the DEP method and from double-oblique to vertical net hauls should improve the precision and reduce costs involved with egg surveys for mackerel, the surveys have already played a significant role in the management of the Western mackerel stock since their initiation in 1977. Catch-at-age analyses based on the commercial catch provide annual estimates of stock size for the Western stock, but could be subject to significant error due to unreported catches or mortality of discarded fish (Lockwood 1988). The egg surveys are independent of the commercial catch data, and help to bring assessments back on course should the "best estimates" of total catch be seriously in error.

Tagging studies provided estimates of stock size and exploitation rate for the North Sea stock during the early years of the fishery, but recoveries eventually declined to the point where there were too few for reliable assessment. As a result, egg surveys have also been the principal method for estimating the size of the North Sea stock since the early 1980s (Lockwood 1988). Egg surveys for mackerel have also been used successfully to obtain estimates of stock size in the Northwest Atlantic (Berrien et al. 1981).

Pacific Herring

When lucrative markets for herring roe were developed in the early 1970s, the number of vessels fishing for herring multiplied rapidly, and a number of "derby" fisheries evolved. In some cases, the entire commercial season for a given spawning stock lasted less than an hour, making it imperative that quotas be as accurate as possible.

A number of survey techniques have been used to provide a rough idea of trends in herring abundance, including aerial surveys of adult spawners in Alaska and surveys of miles of beach covered by intertidal spawning in British Columbia. Acoustic surveys for this species are carried out for some stocks, but the rapid migrations into and out of the spawning areas, together with the need to survey several stocks simultaneously during peak spawning, pose formidable problems. As a result, most management agencies have come to rely heavily on demersal egg surveys for estimates of current spawning biomass.

The sampling problems associated with estimating egg density in areas where spawning deposition is patchy (and varies widely with the spawning substrate) are significant, and the methodology for addressing them is still evolving. Differences in methodology are common from agency to agency, and in some cases cost considerations are such that the estimates for smaller spawning stocks are far more crude than those for the major stocks (Schweigert 1993).

The herring stock in Prince William Sound, Alaska, is an important one, and represents one of the most successful applications of demersal egg survey techniques (Biggs and Funk 1988). During 1988, aerial surveys of adults were made at least daily if weather permitted, and extended throughout the spawning migration (April 18–May 9). Observers charted the location of milt sightings, and these indicated the presence of four major spawning areas. Surveys with small boats were used to define spawning areas further.

The number of eggs deposited in each area was estimated using SCUBA surveys. Transects were placed perpendicular to the shore and selected randomly, and 0.1 m^2 quadrats were sampled at 5 m intervals along each transect. Egg density within each quadrat was estimated visually, and then adjusted using the model described previously (Equation 4.7). The variance of the estimate for total egg deposition in an area was obtained by using standard equations for three-stage sampling (Cochran 1977), modified to take into account the fact that primary units (transects) contained unequal numbers of secondary units (quadrats). The variance for the third stage (diver observation model) was estimated from

$$\mathrm{Var}\,(\hat{Y}_{ijk}) = (e^{2\hat{Y}_{ijk} + s^2})\,(e^{s^2} - 1)$$

where $\hat{Y}_{ijk}$ = predicted egg density for the ith estimate for diver j in vegetation type k

s^2 = mean squared residual error from the visual observation model

Mean fecundity, mean weight, and sex ratio were estimated by sampling purse seine catches made either by the commercial fleet or by test fishing conducted before or after commercial openings. Egg mortality was not estimated directly, and an assumed value was used instead. The variance of the sex ratio was estimated from the binomial distribution, and the variance of the mean fecundity estimate from (Draper and Smith 1981)

$$\mathrm{Var}\,[F(\overline{W})] = s^2\left[\frac{1}{n} + \frac{1}{q} + \frac{(\overline{W} - \overline{W}_{\mathrm{r}})^2}{\sum_i (W_i - \overline{W}_{\mathrm{r}})^2}\right]$$

where $F(\overline{W})$ = fecundity for a female of average weight, as predicted from the fecundity–weight regression

s^2 = residual mean square from the fecundity–weight regression

n = number of females in the regression sample

q = number of females the estimate of mean female weight ($\overline{W}$) is based on

$\overline{W}_r$ = mean female weight in the regression sample

W_i = weight of the ith female used in the regression sample

The variance of the biomass estimate was approximated using the delta method (as in equation 4.10), ignoring some covariance terms. The resulting estimate of biomass for all four of the major spawning areas combined was 42,300 tons, with a CV of 17%. Uncertainty remained as to the true magnitude of egg mortality, since values reported in the literature range from 10 to 52% per day (Palsson 1984; Schweigert 1993), depending on the level of bird predation. Direct estimation of egg loss, using repeated surveys of special index transects, was suggested for subsequent surveys.

5

DIRECT COUNTS

While acoustic surveys of fish populations have their origin in the early twentieth century, and ichthyoplankton surveys in the nineteenth, direct visual counts of fish or fish schools could well date back to ancient times. Modern fisheries scientists have a wide array of technology and methodology at their disposal, however. Direct counts can be carried out from manned submersibles, remotely operated vehicles (ROVs) or aircraft, as well as survey vessels and counting towers. Modern scientists can employ SCUBA, sonar, still cameras, and video cameras to enhance their ability to enumerate target species. A considerable amount of theory has also been developed to aid in quantifying the dimensions of the area being sampled during the counting process.

Most direct counting surveys follow a typical pattern of estimating the density (number or weight/m^2) along randomly or systematically selected sampling transects, and extrapolating the result to the entire survey area to estimate population size ($\hat{P}$) or biomass ($\hat{B}$). Survey designs are often comparable to the discrete transects used in acoustic surveys (Fig. 3-24), and the density observations for each transect ($\overline{D}_j$) are a logical choice for the basic sampling unit. The $\overline{D}_j$ values can be treated as independent observations, suitably weighted to adjust for differences in transect length, and Equations 3.5 and 3.6 used to estimate mean density ($\overline{D}$, number/m^2) and its variance. Most alternatives in survey design and data analysis discussed in the section on acoustic surveys (Chapter 3) also apply to sighting surveys.

Few survey techniques have been subjected to as many violations of availability, vulnerability, and selectivity assumptions as those employing direct counts. The tendency has all too often been to adopt the attitude "Since we can see them, we can count them," and worry later about what fraction of the stock is being counted. As with any direct survey technique, however, the first priority should be to establish that the entire stock of the target species is present in the survey area at some point, and that it is completely vulnerable (visible in this case) to the counting technique being employed.

When many of the fish being enumerated are too deep in the water column to be visible, or can hide under undercut banks and debris in a stream, direct counts should not be undertaken unless it can be established that these represent a fixed fraction of the stock.

Direct counts of schools are particularly suspect in this regard, since the observer can count only those individuals that are aggregated in large, detectable schools during the survey. Any individuals that are scattered or occur in diffuse aggregations are not vulnerable to the survey. During one intensive survey of widow rockfish (*Sebastes entomelas*) schools, for example, Wilkins (1986) estimated the biomass in detectable schools to be 830 mt in an area where the total biomass had been estimated at about 22,000 mt on the basis of catch-age analysis.

A further assumption is that the survey technique neither attracts nor repels the target species, and the scientists employing direct counting techniques have been remarkably resourceful at rationalizing in order to ignore obvious violations of this assumption. The fact remains that direct counting techniques are often applicable only to species whose behavior is such that they either "freeze" when the presence of observers (or their apparatus) is sensed, or are oblivious to it. Natural selection has ensured that few such species persist in the presence of predators, however, and most species will either flee or hide when disturbances are detected. Furthermore, they will frequently detect the presence of the observer long before he detects them. To evaluate avoidance behavior properly then, an independent, noninvasive observation technique is required to observe the behavior of the target species both before and after the observer has passed through the survey area. It should always be kept in mind that avoidance can begin at considerable distances (hundreds of meters) from the observer (see Chapter 1).

COUNTING UNIT

Individuals are usually the basic counting unit, but counts of schools ("pods" or "groups" in the case of some marine mammals) are made for several species. While the counting techniques are identical in both cases, the goal is to estimate individual density (D, or number/m^2) in one case, and school density (D_s, schools/m^2) in the other. If a population estimate is required for schooling species, additional information is required on the number of individuals within an aggregation. Total population size can then be determined from

$$\hat{P} = A\overline{D}_s\overline{S} \tag{5.1}$$

where A = total survey area (m^2)
$\overline{D}_s$ = number of schools/m^2
$\overline{S}$ = mean school size (number of fish per school)

An estimate of biomass ($\hat{B}$) can be obtained by expressing D_s and $\bar{S}$ in terms of weight rather than number.

In the case of fish schools, the estimation of $\bar{S}$ requires additional field work on an extensive scale, since both the size of the schools and their packing density (number or weight/m^3) vary widely. A further problem in using schools as the observation unit lies in the fact that they are dynamic in size and shape, breaking up and reaggregating over very short time scales (Gunderson et al. 1981; Wilkins 1986; Buerkle and Stephenson 1991). They can also be highly mobile (Buerkle and Stephenson 1991), posing problems of double or missed counts when the survey progresses slowly in relation to the movement of the target species. One school group of sardines was observed to move 27 nm in 24 hours, averaging approximately 1 knot (Cram and Hampton 1976).

Fish schools and aggregations of marine mammals are typically quite patchy in their distribution, and it is usually necessary to cover large expanses of empty water before locating them. This is a familiar problem to the commercial fishermen who exploit these species, and often necessitates the use of "spotter" planes to locate fishable concentrations. Such spotter planes are employed in anchovy, sardine, menhaden, herring, mackerel, and tuna fisheries around the world.

In the late 1970s and early 1980s several attempts were made to conduct summer acoustic surveys for herring off southwest Nova Scotia (Buerkle and Stephenson 1991), but were unsuccessful because the herring aggregated in very mobile schools, in restricted locations that varied from night to night. During the daytime, they disappeared. The pattern of the commercial fleet was to leave port at dusk, fan out over a large area searching for the herring with acoustic equipment, and converge when fish were located. During surveys, as many as 128 vessels were seen fishing in an area of less than 3 km^2, and no herring could be detected around the outside of that area. During one survey in 1981, 93% of the total biomass was located in one small 30 km^2 area. In 1984 and 1985, the surveys were successful in mapping large concentrations of herring in one 340-km^2 area, while no herring were found in the 800,000-km^2 area of the historical fishery.

When the target species is distributed in this manner, with enormous distances separating dense aggregations that are highly localized, it may be necessary to adopt a two-stage survey design. Aircraft or long-range sonar can be used to locate areas of aggregation, followed by detailed estimates of the number or biomass in each aggregation using supplementary surface vessels. This can be costly, however, and one alternative is to rely on the commercial fishing fleet to locate all important aggregations. However, if the latter course is followed, doubt will always remain as to whether all important aggregations have been censused. This problem is not confined to surveys of schooling fish, since Myers and Bowen (1989) concluded that failure to detect all whelping concentrations was likely to be the largest source of error in aerial surveys of harp seals.

DETERMINING THE AREA SAMPLED

Strip Transects

In cases where strip transects are employed, the boundaries of the sampling area are established prior to the survey, and only individuals that lie within these boundaries are counted. It is assumed that all individuals lying within this strip remain there as the observer passes, and are detected by the observer. This is a natural design where the target species is mobile or difficult to detect, and the observer must keep his eye on the field of observation at all times in order to avoid missing an observation or double counting. Under these circumstances, or when the observer's progress through the sampling area is rapid (most aerial surveys), there is little time to collect any data other than counts, and measurements of sighting distances, sighting angles etc., are not possible.

Aerial Surveys

During aerial surveys, the width of the observer's sampling strip depends on altitude, as well as the deflection angles (α, β) from the horizon line that are selected (Fig. 5-1). For fixed wing aircraft, the observation area is limited by the size of the windows, presence of wheels, and other obstructions. The width of the sampling strip can be given by

$$W = a(\cot \alpha - \cot \beta)$$

where a = altitude.

In the case of helicopters (Fig. 5-1), the observer can be positioned so that visibility on either side of the aircraft is good, and the width of the sampling strip is given by

$$W = 2a \cot \alpha$$

The deflection angles can be checked in flight using a range finder or sighting board (Fig. 5-2). Marks or streamers can be attached to the struts and windows to aid in demarcating the sampling area, and the accuracy of the estimate of W checked by overflying a reference strip that has previously been measured and marked on the ground. Breaks should be allowed at intervals either within or at the end of aerial survey transects to allow the observer to check navigation, change position, and relax briefly before concentrating on the next series of counts.

The altitude chosen can vary from 100 to 500 m, depending on the topography and target species. Operating at a lower altitude can startle the target species (especially when helicopters are employed), while higher altitudes are vulnerable to the effects of cloud cover and reduced ceilings. Viewing conditions vary with the angle of the sun, and are usually best when the sun is

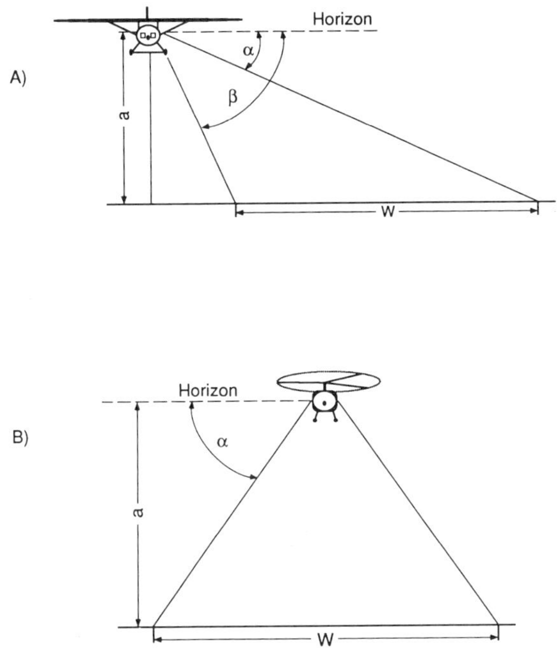

Figure 5-1. Width (*W*) of the sampling strip when observations are made at altitude *a* from (A) a fixed wing aircraft and (B) a helicopter.

at its highest, minimizing both glare and shadows. Biological considerations often dictate the time when flights are conducted, however. In the case of surveys for pinnipeds in the Antarctic for example, there is diel cycle in the number of individuals "hauled out" on the ice where they are visible. The highest densities of crabeater seals are regularly observed between 1130 and 1500 (Erickson et al. 1989), and it is critical that either all sampling is con-

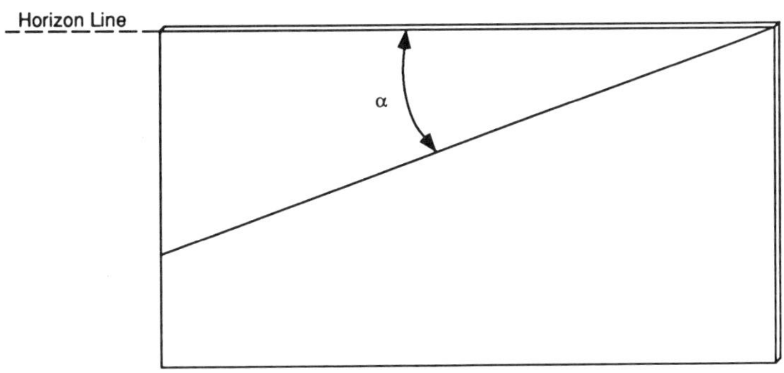

Figure 5-2. Schematic drawing of a range-finding board. The observer aligns the top of the board with the horizon, and sights down the line determined by angle α to determine transect width.

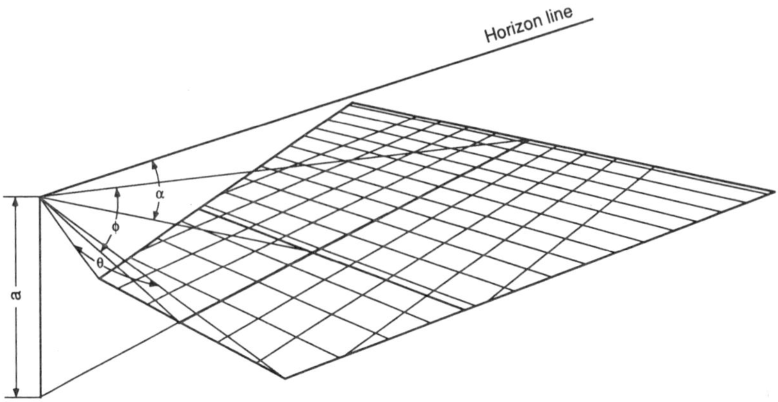

Figure 5-3. Geometry of a perspective grid, where α is the inclination of the camera axis from the horizontal, *a* is the altitude of the camera lens above the observation field, and θ and φ are the camera's horizontal and vertical angles of view.

ducted within this time window or a model of the diel cycle be developed to standardize all counts to the peak haulout period. Even then, an appreciable number (about 17% based on data from individuals that had been fitted with time–depth recorders) of seals do not haul out during a given day, and are unavailable to the census (Erickson and Hanson 1990).

Aerial photography may offer some advantages over direct visual counting, particularly where data on school sizes or number of individuals per group are required. The choice of camera and film used depends on the target species, terrain to be photographed, and type of aircraft. Counts and size measurements are often difficult to obtain in the case of oblique photographs taken with hand-held 35-mm cameras. Even when the angles employed are fixed, spatial scales from oblique photographs change substantially with distance from the bottom of the image (Fig. 5-3).

Large format vertical photographs are easiest to interpret, and minimize the problems of marking out areas that have been photographed twice. Proper interpretation of aerial photographs requires experience, and quantitative experiments to measure consistency should be an integral part of the analysis of photographic aerial surveys (Myers and Bowen 1989). Hiby et al. (1987) describe a relatively inexpensive rocking camera mount for paired 70-mm cameras that compensates for the movement of the airplane. Hiby also describes a method for focusing the camera to achieve maximum resolution. However, large format cameras are heavy, and may be impractical in instances where the superior resolution, greater depth of field, and binocular vision provided by the human eye are required for identification of species or age groups.

In some situations, the turbulence associated with fish schools causes bioluminescence, and aerial censusing is best accomplished at night. Cram and Hampton (1976) used a low-light television camera with a wide angle lens and a videotape storage system to conduct nighttime surveys of sardine schools off Africa, allowing detailed measurements of each school that was detected. Independent observations from research vessels showed that at night, fewer than 7% of the schools were deeper than 8 m, the approximate limit of visibility under normal conditions. Anchovy schools off California are also shallowest at night, and can be detected using low-light television (Squire 1978).

SCUBA Surveys

SCUBA allows a diver to work for extended periods in shallow water (e.g., about 1 hour in less than 20 m), and allows detailed *in situ* observations of fish behavior and habitat use. Quantitative SCUBA surveys are common, and have been conducted in a variety of shallow-water habitats ranging from marine reefs to freshwater streams.

The transect lines for SCUBA surveys can be demarcated with a series of lead-cored ropes or ropes fixed to the bottom with weights or pitons. The sampling width must be restricted to that area in which all individuals can be detected and enumerated, typically ranging from 2 to 5 m depending on observation conditions, and the behavior, size, and conspicuousness of the target species. When the target species is abundant, the accuracy of the census can be increased by conducting a greater number of short (20–25 m) rather than a few very long (100–200 m) transects (Brock 1982). The survey area should be subdivided into several strata (Chapter 1) whenever different habitat types exist or when the densities or size composition of the target species vary spatially.

Counts are often recorded using a pencil and waterproof paper or plastic slates, scrolls, or wrist cuffs (Helfman 1983). Recording the data with an underwater tape recorder allows observers to keep their attention focused on the observation field at all times, however, and can result in greater accuracy (Greene and Alevizon 1989). SCUBA surveys are sometimes conducted using a video camera to record all observations, but controlled experiments have shown that this is less accurate than direct visual counting (Greene and Alevizon 1989). This was attributed to the narrow field of view, poor light gathering ability, and poor resolution of the traditional video cameras used during the comparisons. However, the use of high-resolution video cameras could give better results. S-VHS and Hi-8 recorders offer 400 horizontal lines of resolution, while VHS offers a resolution of only 230 lines.

It is difficult to accurately identify and count large numbers of species and individuals simultaneously, and complex fish communities (such as those inhabiting tropical reefs) should be subdivided into discrete groups. Each subgroup (a single species or group of species) can then be censused in a separate survey. This allows the observer to form a more specific "search image" of species with similar behavior patterns, appearance, or distribution, and improves the overall accuracy of the survey (Greene and Alevizon 1989).

SCUBA surveys are usually confined to daylight hours, owing to the difficulty and hazards of night diving. In addition, the illumination required at night adds a further cue as to the observer's presence. Many of the problems inherent in night diving also apply to turbid waters.

In one of the few experiments where the assumptions of availability and vulnerability of the target species were examined rigorously, Brock (1982) visually censused a 1500 m^2 patch of reef using standard strip survey techniques, and subsequently encircled the area with a net and applied rotenone. The results showed that SCUBA surveys are accurate only when censusing large, conspicuous, diurnally exposed species that are resident in the survey area. Migratory species are poorly censused, as are cryptic (well-camouflaged or secretive) species. At higher abundances, the error associated with visual estimates may be large, and underestimation results. Avoidance of or attraction to the observer will also result in serious bias for some species.

To a large degree then, the accuracy of a SCUBA survey is dependent on the characteristics of the target species, and the investigator's knowledge of their behavior.

Snorkeling Surveys

Snorkeling surveys are appropriate where water clarity exceeds water depth, and are employed commonly in freshwater streams. The width of the sampling strip can be set by spacing a team of divers at fixed intervals along a single length of PVC pipe, and assigning a specific counting lane to each (Schill and Griffith 1984). Snorkeling surveys should not be undertaken in shallow water (less than about 0.75 m deep), or where visibility is less than about 3 m. SCUBA is required in large, deep streams, and may often be preferable to snorkeling in smaller streams (Gardiner 1984).

Snorkeling counts can be relatively precise, with 95% confidence limits of ±25% or less (Schill and Griffith 1984; Hankin and Reeves 1988), but can be expected to underestimate the true population. Behavioral observations during one survey revealed that some fish moved downstream in advance of the survey team, initiating avoidance responses when the team was up to 91 m away (Northcote and Wilkie 1963). By netting off a section of stream that had previously been surveyed, and applying rotenone, it was determined that the counts seriously underestimated the number of fish actually present. The mean diver counts from 12 replicates of this section underestimated the number recovered after rotenone poisoning by 36% in the case of mountain whitefish (*Prosopium williamsoni*), 41% in the case of rainbow trout (*Oncorhynchus mykiss*), and 50% in the case of largescale suckers (*Catostomus macrocheilus*).

Further problems arise when the fish being surveyed occupy areas that are too shallow to be censused by snorkeling, a situation that often limits the application of snorkeling surveys to 0+ salmonids. In many situations, the target species may also be able to hide in the bottom substrate, woody debris, or undercut banks. Hankin and Reeves (1988) reported that steelhead trout

(*Oncorhynchus mykiss*) were more difficult to survey than coho salmon (*Oncorhynchus kisutch*), due to the fact that they were more closely associated with the bottom, and were quicker to move to cover when disturbed. Because of this difference in behavioral response to observers, the protocol used was to scan the bottom immediately on entering a pool, and count trout before salmon.

Hankin and Reeves (1988) corrected for the bias in snorkeling surveys by calibrating visual estimates against more accurate estimates obtained through depletion experiments (see Examples of Successful Applications).

Underwater Camera Sleds

Camera sleds have been used extensively in benthic ecology work, but their utility in quantitative fisheries surveys has been restricted by the limited number of species that inhabit areas of smooth bottom topography, and whose behavior is not influenced by the noise and light that the sleds generate. This situation is exacerbated by the fact that the severe attenuation of light by water often confines the viewing area to the immediate vicinity of the camera.

When still cameras are employed, a series of nonoverlapping photographs usually results, with several images being discarded due to problems with poor visibility, sediment clouds, and tilting or rolling of the sled. Continuous coverage of the observation zone is not possible, and the photographs must be treated as a series of quadrat samples along a sampling strip.

A typical 35-mm sea camera consists of a chassis, film mechanism with electronic controller, and lens, all enclosed in a pressure housing. The use of larger format cameras (e.g., 70 mm) will result in recording greater detail, at the cost of an increase in camera size and a decrease in the ease of processing. A common lens used in deep-sea cameras is a 28-mm water-corrected lens (Butler et al. 1991). For many applications, the widest angle lens is desirable, since it reduces the distance between the subject and the lens, thereby minimizing the loss of light due to attenuation and the loss of resolution due to backscattering of light from particulate matter in the water. Separation of the camera and lighting system will also reduce the backscattering problem. Camera sleds are typically constructed out of welded steel for deep sea work, and can weigh 500–650 kg. Lighter models, constructed out of PVC pipe have been described for other applications (Bergstedt and Anderson 1990).

The scale within an oblique photograph can be determined empirically by taking calibration photographs of an underwater grid that has been marked out or placed on the bottom (Uzmann et al. 1977; Bergstedt and Anderson 1990). These photographs can then be used to draw a template on a transparency, which can then be overlaid on photographs taken during the survey.

Alternatively, the grid can be deduced from photogrammetric principles (Wakefield and Genin 1987) knowing (Fig. 5-3):

1. The height of the camera off the bottom,
2. The inclination angle (α) of the camera's optical axis, and
3. The camera's horizontal (θ) and vertical (ϕ) angles of view.

The geometry of the grid in Figure 5-3 can be derived from these parameters, and used to calculate distances along both the horizontal (parallel to the base line) and vertical scales and to draw a grid on a template. A computer program for determining sizes and distances can also be developed, and interfaced with an electronic digitizing system (Wakefield and Smith 1990). It must be assumed that all objects being measured are directly on the bottom.

When the sled is towed over uneven terrain, changes in the inclination of the camera can result in error that increases almost exponentially with distance from the bottom or toward the edges of the image (Wakefield and Genin 1987). This error is particularly serious for cameras with small angles of inclination (α), and the design of the camera system should consider the tradeoff between the size of the area covered in each photograph and the decrease in the proportion of the photograph in which measurements of distances and sizes are accurate.

Determination of the effective sampling area depends on a variety of factors, including the degree of backscattering from the strobe lights, the reflective properties of the target species, and the relief of the sea floor. As a result the sampling area differs between species, and even for different size groups within species. The effective sampling area should be determined by examining data on sighting frequency versus distance (see Fig. 5-5), both in front of the sled and to the side. Analysis of sighting data for Dover sole (*Microstomus pacificus*) and thornyheads (*Sebastolobus* spp.) indicated that the area over which the probability of detection was constant extended only about 180 cm in front of the sled, and about 70 cm to the side (Butler et al. 1991). The sighting data for Dover sole showed a peak at 50 cm rather than decreasing monotonically with distance from the transect line, suggesting movement in response to the camera sled.

Manned Submersibles

The design and configuration of manned submersibles are quite variable, but most accommodate a pilot and at least one observer within a pressure hull. In one design (Deep Rover), the observer does the piloting.

There are typically two means of directly viewing the outside environment (Butler et al. 1991), either small circular view ports (11 cm diameter in Alvin, Pisces, or Turtle) or an entire pressure hull manufactured from transparent material (Johnson Sea Link, Deep Rover). In submersibles with small view ports, the observer's field of view is normally quite limited, and the acrylic view ports present a magnified image. In those submersibles with acrylic pressure hulls, the pilot and observer have a wide field of view, but the apparent size of objects is somewhat reduced. Powerful lights are required to illuminate the observation area, which is consequently limited to the 2.5–11 m immediately adjacent to the submersible (Uzman et al. 1977; Pearcy et al. 1989; Matlock et al. 1991), depending on water clarity and the visibility of the target species.

Avoidance of the submersible and attraction to it are common phenomena, and should be evaluated carefully prior to undertaking quantitative surveys. Comparison of trawl catches and submersible observations near Georges Bank showed that herring, mackerel, and butterfish were able to completely avoid both camera sleds and submersibles (Uzman et al. 1977). Canary rockfish (*Sebastes pinniger*) were seldom seen during submersible dives at Heceta Bank, Oregon, despite the fact that this species made up 88% of the total weight in trawl catches from the same area (Pearcy et al. 1989). In contrast, yelloweye rockfish (*Sebastes ruberrimus*), greater amberjack (*Seriola dumerli*), and black jack (*Caranx lugubris*) are attracted to submersibles, and actively follow them about (Carlson and Straty 1981; Ralston et al. 1986). While attraction is easy to detect from observations made from the submersible, remote observation (e.g., using echosounders, multibeam sonar or acoustic tags) is required in order to be certain that avoidance responses are not occurring beyond visual range.

A common problem in submersible (and SCUBA) work is the difficulty of species identification. Many species cannot be differentiated on the basis of visual observations, often making trawl surveys a necessary adjunct to submersible observations. This combination of techniques has the added advantage of combining detailed observations of habitat use from the submersible with broad geographic coverage from the trawl survey. This allows the assessment scientists to be certain that area swept observations from the trawl are being extrapolated over the appropriate area. Recent observations from a submersible indicate that large concentrations of adult Pacific ocean perch (*Sebastes alutus*) do not inhabit extremely rough bottom (Kreiger 1993) and that previous extrapolations of trawl survey density estimates to "untrawlable" bottom may result in overestimates of population size. These observations substantiate previous results obtained through gill net sampling (Matthews and Richards 1991) in untrawlable areas.

Submersible surveys have followed a variety of designs. Strip transects were employed by Uzman et al. (1977), who suspended a chain from a wooden yardarm to delineate a sampling path width of 2.4 m within which all fish and invertebrates seen were counted. Pearcy et al. (1989) used two fixed points on the submersible's bumper to define a sampling strip that was 3.5 m wide, when altitude was maintained at 2 m. Altitude was determined by using a length of chain that had been marked in decimeters, and suspended from the submersible. A fiberglass "T"-shaped rod protruding from the bow with black and white bands at 10 cm intervals, and a similarly marked bar attached to the bottom of the chain were used to estimate fish lengths. A "slurp gun" and rotenone dispenser was used to collect specimens and verify species identification. External lighting consisted of two 500-W and two 150-W iodide lamps. A 35-mm camera and underwater television camera were used on every dive to assist in identification of fishes after the dives.

Rolling and yawing of the submersible often make it difficult to maintain a constant altitude and attitude when following a transect, particularly when currents are strong or when the submersible must frequently maneuver around rocks or pinnacles. Ideally, these variables, together with time and position data should all be incorporated into a single data logging system and used to correct estimates of the effective observation area after the survey is completed. The development of laser systems that project a visible grid onto the sea floor may provide an alternative approach. Two lasers allow scaling of organisms and three lasers define the plane of view with fixed focal length lenses. The addition of a fourth laser at an angle will provide this information for zoom lenses at any focal length (Butler et al. 1991).

Remotely Operated Vehicles

The expense involved with the use of manned submersibles has limited their application in fisheries survey work. Remotely operated vehicles (ROVs) are less expensive to operate, although the substitution of video cameras for the human eye results in some costs. The field of view is sharply reduced, and the high resolution, depth of field, and sensitivity of the human eye must be replaced with a series of lenses and recording systems. During controlled experiments with known populations of fish, visual SCUBA surveys clearly outperformed surveys made with a Deepsea Systems MiniRover (Greene and Alevizon 1989). Nevertheless, ROV video transects can be conducted well below SCUBA depth, and can potentially be more cost-effective than manned submersibles given the extensive areas that must be covered during most fisheries surveys. Advances in ROV design, high-resolution video systems, and low-light video technology should improve the effectiveness of ROV surveys in the future.

The ROVs used in survey work require high-resolution video cameras, adequate lighting, and minimal loss of picture clarity in transmission. Only fiberoptic tethers meet the last requirement, and a tether management system is necessary at depths below about 400 m (Butler et al. 1991). The ROV must monitor and record information on altitude, camera angle, zoom, and focus to allow quantitative estimation of the size of the field of view. Laser spots can provide a convenient way to calibrate the video system when they are available.

Line Transects

Line Transect Theory

When logistics allow, it is sometimes preferable to obtain data on sighting distances for each individual (or school) counted. This allows the observer to do away with strict a priori definition of the sampling strip, fit the data on sighting distances to an appropriate model, and use the model to define effective sampling width. This extends the observation area to as wide an area as possible, and allows the observer to obtain information on almost all individuals or schools observed.

The line transect method is based on the fact that the probability of sighting a given individual or school varies with distance from the observer. Burnham et al. (1980) defined the probability density function for the perpendicular distance data $f(x)$ as

$$f(x) = g(x)/a$$

where $g(x)$ = probability that an individual (or school) will be detected given that it is at perpendicular distance x from the observer (detection function), and

$$a = \int_{x=0}^{x_t} g(x)dx,$$

where x_t is the truncation distance beyond which all observations are discarded. Normally, 1–3% of the data should be truncated (Burnham et al. 1980).

Thus, $f(x)$ is simply $g(x)$ scaled so as to integrate to 1. If it can be assumed that all individuals on the transect center line are sighted, then $g(0) = 1$, and $f(0) = 1/a$.

Once $f(0)$ has been estimated, Burnham et al. (1980) show that the effective sampling width (W) can be estimated from

$$W = 2/\hat{f}(0)$$

Hence the quantity $1/\hat{f}(0)$, or a, can be conceptualized as one-half the effective sampling strip. If individual fish or schools on both sides of the transect are counted, density can be estimated from

$$\hat{D} = \frac{n\hat{f}(0)}{2L}$$

where n = number of individual fish (or schools) counted
$\hat{D}$ = number of individuals /m^2
L = length of transect (m)

The same relationship will hold when the observer is counting schools rather than individuals, and estimating school density ($\hat{D}_s$, number of schools per m^2).

Data on perpendicular sighting distances can be obtained either by direct measurement or by measuring radial distance and sighting angle (R and θ in Fig. 5-4) and calculating perpendicular distance from $x = R \sin \theta$. The data collected on sightings can be fitted to a number of models (Burnham et al.

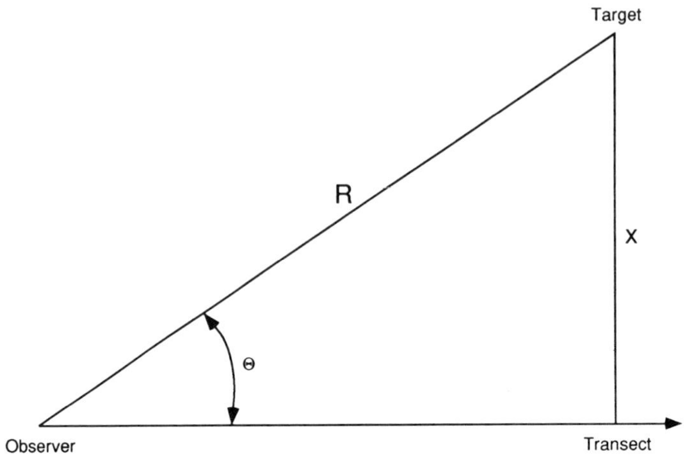

Figure 5-4. Geometry of a line transect survey. The observer can either estimate perpendicular distance to the target (x) directly, or calculate it using estimates of the range (R) and sighting angle (θ) to the target.

1980; Quinn and Gallucci 1980; Buckland 1985), all of which depend on the following assumptions:

1. Transect lines have been placed randomly with respect to the target objects.
2. Target objects on the transect line will always be sighted [i.e., $g(0) = 1$].
3. Target objects are sighted at the position they occupied prior to the approach of the observer.
4. Sightings are independent events.
5. Perpendicular distances to target objects are measured precisely, particularly near the transect line.
6. No target object is counted more than once.

Numerous parametric and nonparametric estimators have been developed to analyze sighting data. Parametric models (Quinn and Gallucci 1980) assume a fixed shape for the detection function a priori, and include the exponential, logistic, half-normal and hazard rate models. Models that do not assume a shape a priori are termed nonparametric. Examples of these include the Fourier series and Hermite polynomials. Burnham et al. (1980) concluded that the nonparametric Fourier model is a robust, flexible estimator for $f(0)$.

Sighting data for schools present a special case, since the sighting function often varies with school size. Breaking the data down by school size can result in small sample sizes for some size groups, and Burnham et al. (1980) suggest a minimum sample size of 40 objects for line transect estimates of density. Fortunately, Quinn (1985) has shown that there is theoretical justification for pooling data for several size classes of schools, since the sighting model that results from the pooled model is self-weighted by the relative abundances of school size.

The Fourier model is based on a Fourier series expansion of the probability density function $f(x)$. The estimator for $f(0)$ (Burnham et al. 1980) is

$$\hat{f}(0) = \frac{1}{x_t} + \sum_{k=1}^{m} a_k$$

where x_t = truncation distance

$a_k = \frac{2}{nx_t}\left[\sum_{i=1}^{n} \cos\left(\frac{k\pi x_i}{x_t}\right)\right]$

x_i = perpendicular distance off transect for the ith individual or school observed

k = term number = 1, 2, 3, . . . , m

n = number of individuals or schools observed

Bias in the estimation of $f(0)$ is reduced by increasing the number of terms employed, but this reduction comes at the cost of reduced precision. Burnham et al. (1980) present a stopping rule to choose the appropriate number of terms (m) in the Fourier series. Buckland (1985) presents an alternative variable-term model, using Hermite polynomials, which is less sensitive to the number of terms selected than the Fourier series estimate.

Hayes and Buckland (1983) developed a parametric hazard rate model, which led to a detection function of the form

$$g(x) = 1 - \exp[-cx^{-(b-1)}], \qquad b > 2$$

where b and c are constants estimated by using an iterative maximum likelihood procedure.

Since the derivative of $g(x) = 0$ at $x = 0$, the hazard rate model possesses a "shoulder" (Fig. 5-5). Burnham et al. (1980) concluded that a shoulder in the detection function is a logical condition suggested by the sighting process itself, and its existence is an important criterion in selecting a model. When the data themselves fail to show a shoulder, the target species could be avoiding or attracted to the observer, resulting in serious violations of assumptions.

Biased estimates of abundance will result if movement is not corrected for (Turnock and Quinn 1991), a procedure that requires auxiliary studies of the behavior of the target species in response to the observers. Auxiliary studies such as acoustic tag tracking, aircraft or helicopter observations in advance of the survey vessel, or observations with long-range multibeam sonar are a good idea in any event, in order to examine the validity of the assumption that no movement occurs in response to the observer. It should be borne in

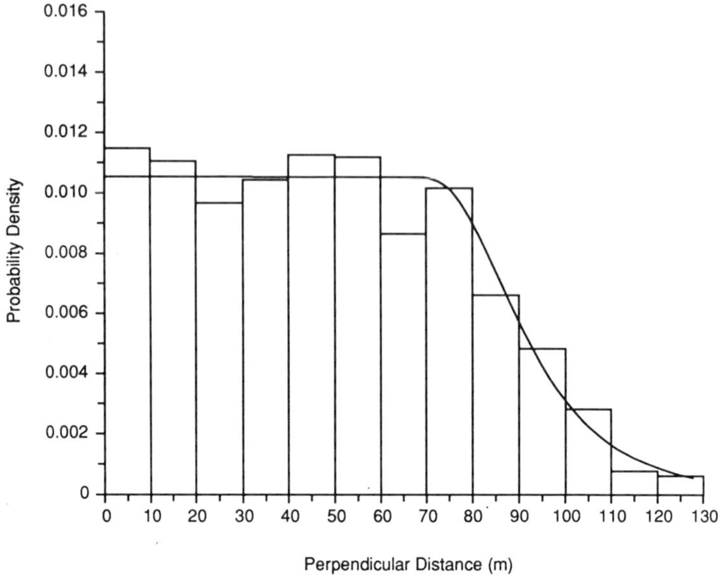

Figure 5-5. A hazard rate model (solid line) fitted to grouped data on perpendicular sighting distances (vertical bars).

mind that even when the survey targets are moving randomly, bias may result if their speed is significant with respect to the speed of the observer (Turnock and Quinn 1991).

Buckland (1985) compared the performance of several sighting models, using a variety of data sets gathered in the field. The analysis was carried out on data that had been grouped into equal intervals of sighting distance (Fig. 5-5), since this minimizes the impact of inaccuracies in measurement errors on the estimation of $f(0)$ (Burnham et al. 1980). Maximum likelihood estimates and large-sample variances were obtained using the multinomial method described by Burnham et al. (1980), and the results indicated that the hazard rate model provided somewhat better fits than the Fourier series model. All models provided similar estimates of $f(0)$ when the data set exhibited a shoulder.

Surface Vessels

Line transect surveys from research vessels have been employed in a number of surveys of marine mammals (whales, dolphins etc.), and have found limited application in sonar surveys of schooling fish. There are often problems with vessel avoidance, particularly when the target species is capable of attaining higher speeds than the survey vessel, and detecting its presence at considerable distances. Problems with availability are also significant, since the target species must be on the surface in order to be detected visually, and aggregated in schools in order to be detected with a sonar.

Violations of critical assumptions often result in estimates that can provide only a relative measure of density. Even then it must be assumed that variations in availability, avoidance, and vulnerability are such that any bias in the density estimate is consistent from survey to survey, an unlikely occurrence for many species (Chapter 1).

Estimating Mean School Size

MARINE MAMMALS. In many applications employing surface vessels, counts are made of schools, aggregations, pods etc. rather than individual fish or mammals. As mentioned previously, mean school size (numbers or biomass) must be determined in order to convert estimates of school density to population estimates. This is most readily accomplished in the case of marine mammals, where it is feasible to count the number of individuals in each group. Since many of these species often spend far more time submerged than at the surface (Hiby and Hammond 1989), counting each individual in a given school a single time can be extremely difficult. Missed counts and double counts can be common when there are several individuals in a given school, and they are coming to the surface to breathe at irregular intervals.

A number of whale and porpoise studies have shown that large schools are more likely to be detected at greater distances than small ones, and this must be taken into consideration when estimating mean school size. Estimating the number of individuals per group on the basis of sightings near the transect line will yield biased results. Quinn (1985) found that mean perpendicular sighting distance for Minke whales increased as the log of school size, i.e., $h(S_i) = a + b\ln(S_i)$. He then developed a weighted estimator of mean school size ($\overline{S}$):

$$\overline{S} = \frac{\sum_i n_i S_i [h(S_i)]^{-1}}{\sum_i n_i [h(S_i)]^{-1}}$$

where n_i = number of observations for school size class i (S_i). If the estimate of total population size ($\hat{P}$) is given by

$$\hat{P} = A\overline{D}_s\overline{S}$$

where A = total survey area (m^2)

$\overline{D}_s$ = estimated number of schools/m^2

the variance of this estimate is given by (Goodman 1960)

$$\mathrm{Var}(\hat{P}) = A^2\left[\overline{D}_s^{\,2}\,\mathrm{Var}(\overline{S}) + \overline{S}^2\,\mathrm{Var}(\overline{D}_s) - \mathrm{Var}(\overline{S})\mathrm{Var}(\overline{D}_s)\right]$$

Estimates for $\mathrm{Var}(\overline{D}_s)$ and $\mathrm{Var}(\overline{S})$ can be obtained in a number of ways (Burnham et al. 1980, Buckland 1985, Quinn 1985), and can then be used to estimate the variance of $\hat{P}$.

As an alternative, bootstrapping can encompass a variety of different approaches and complications in methodology, and provide direct estimates of $\mathrm{Var}(\hat{P})$. Holt and Sexton (1990) obtained a bootstrap estimate of $\mathrm{Var}(\hat{P})$ by subdividing line transect observations for dolphins into a number of time intervals during which sighting conditions were consistent. These intervals were then sampled randomly (with replacement) until a searching effort equivalent to that of the survey had been attained. For each set of survey data, school density (D_s), school size ($\overline{S}$), and species composition of the dolphin schools were estimated. The process was repeated 100 times, with each iteration providing an estimate of P. The hazard rate model was used to estimate D_s, since it does not require subjective selection of the number of terms in the model, and is consequently more readily bootstrapped than the Fourier series model. $\mathrm{Var}(\hat{P})$ was determined from the variance of the bootstrapped estimates of P.

SONAR SURVEYS OF FISH SCHOOLS. Schools of herring and sardines have often been observed to avoid survey vessels (Cram and Hampton 1976; Misund 1990), so that attempts to estimate their biomass with an echosounder can lead to biased estimates of density. Misund and Aglen (1992) found that 35% of the schools detected with a multibeam sonar and aimed at by the vessel were not recorded by the hull-mounted echosounder.

Sonar observations of fish schools can be made at considerable range, and consequently offer one means of counting schools, measuring their dimensions, and observing school behavior before they have responded to the presence of the survey vessel. Sonar estimates of school density (D_s, number of schools/m^2) have consequently been obtained for a number of species using both strip transect and line transect surveys.

Single-beam transducers are usually tilted away from the perpendicular when using sonar, but the basic principles of sound propagation and echo location are the same as outlined for echosounders in Chapter 3. Schools are pin-pointed in range and bearing by echo-ranging and knowledge of the angles over which sound is being transmitted and received. The results are easiest to interpret when they are presented in a two-dimensional display (television or CRT) of the range and bearing of all targets. Some manufacturers offer displays that store and update the results from each sonar scan, and employ a computer to process the information provided by the echo returns. These systems can display the path followed by the survey vessel, the location of the vessel in reference to the school, and auxiliary information on school depth, school speed, and the direction that the school is traveling.

The sonar display can be videotaped for later analysis, and school counts, perpendicular distance measurements (Fig. 5-4), and measurements of school dimensions obtained once ashore. Some preliminary calibration work should be undertaken to obtain scaling factors (s) that will relate distances measured

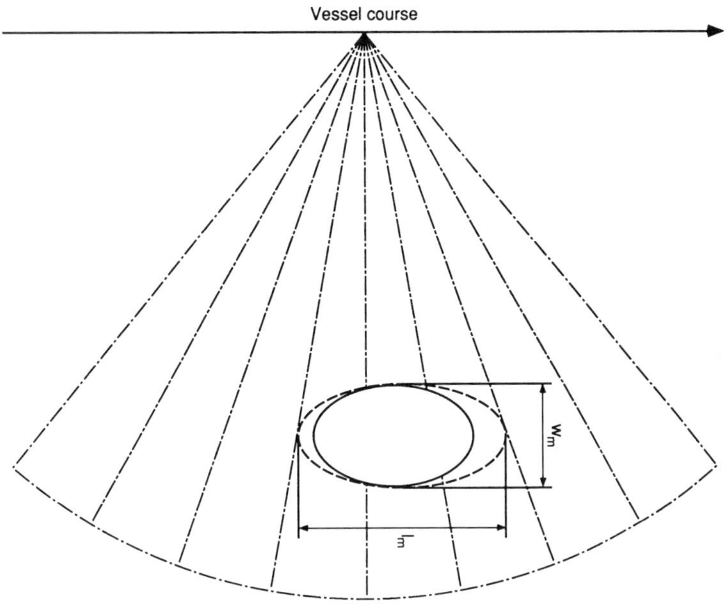

Figure 5-6. Relation between true dimensions for a school (solid line) and the apparent (dashed line) dimensions (l_m = measured length, w_m = measured width) that result from the limited resolution of a multibeam sonar.

on the sonar display to actual distances in the field. Measurements of distances and school width in the direction perpendicular to the vessel's course are the most accurate (Fig. 5-6), being influenced only by range discrimination. The amount of error will depend on the pulse duration being employed, i.e.;

$$w = sw_m - c\tau/2$$

where s = scaling factor which adjusts distances measured on the sonar display (mm) to true distances (m)
w = school width (m)
w_m = school width measurement on sonar display (mm)
τ = sonar pulse duration (milliseconds, ms)
c = speed of sound in water (m/s)

Measurements of school length are artificially inflated due to limitations in lateral resolution by the sonar (Fig. 5-6), and must be corrected for using

$$l = sl_m - 2R\tan(\theta/2)$$

where l = school length (m)
l_m = measurement of school length from sonar display (mm)
R = Range (distance) from vessel to center of school
θ = effective beam angle of sonar (degrees)

In practice it is not possible to determine effective beam angle (θ) without careful calibration, and detailed knowledge of school target strengths (Chapter 3). School target strength data are seldom available, however, and can be expected to vary widely due to differences in school size and packing density (number /m^3). The aspect that the schools present to the vessel changes with depth and range from the vessel, and the target strength of both individual fish and schools will vary accordingly. In most instances, the survey scientist can make only a rough estimate of what the effective beam angle actually is. As a result of this uncertainty, Hewitt et al. (1976) did not even measure l during their surveys of anchovy schools, and estimated school area by assuming that the schools were circular, with diameter $= w$. This assumption was not borne out by later aerial observations of anchovy school shapes, however (Squire 1978).

Accurate measurements of school height (or thickness) cannot be obtained using most sonar systems, and some investigators estimate this parameter by measuring those schools that the vessel passes directly over using an echo-sounder. It is always possible that these measurements are not representative of the schools detected by the sonar, however, owing to avoidance of the vessel, or alterations in school dimensions in response to the survey vessel. As a result, estimating the volume of a school is even more problematic than estimating its area.

The exact range over which schools can be detected by sonar depends on school target strength as well as the source level, frequency, and pulse length employed by the system. Nominal detection range for a school with a 0 dB target strength is up to 1300 m for a SIMRAD SX200 transmitting at a frequency of 57 kHz (using pulse durations of up to 25 ms), and up to 3000 m for a SIMRAD SR240 operating at 24 kHz. When used in quantitative survey work, however, the sonar must be able to detect even the smallest schools throughout the observation range, greatly limiting the distances over which sonar can be used to count schools and measure their dimensions. Further restrictions on effective operating range result from the fact that range resolution is poor when using longer pulse lengths, making measurements of school width relatively imprecise ($c\tau/2 = 18.8$ m for a pulse duration of 25 ms), and that resolution of the height and length of schools deteriorates rapidly with range due to the spreading of the acoustic wave. Refraction of the sonar beam (Fig. 3-2) can result in further problems as range increases. As a result, the effective working range of most sonar systems that have been employed in surveys has ranged from 90 to 450 m (Hewitt et al. 1976; Gunderson et al. 1981; Misund and Aglen 1992).

Sonar systems are widely used by commercial fishermen to locate schools, and a variety of designs are available. Single beam systems typically depend on a transducer that produces a beam about 5–13° wide. Broad area coverage is achieved by placing the transducer in a mounting that can be tilted (vertical plane) and rotated (horizontal plane) mechanically. Scanning in the horizontal plane can usually be preprogrammed by the operator to scan selected sectors

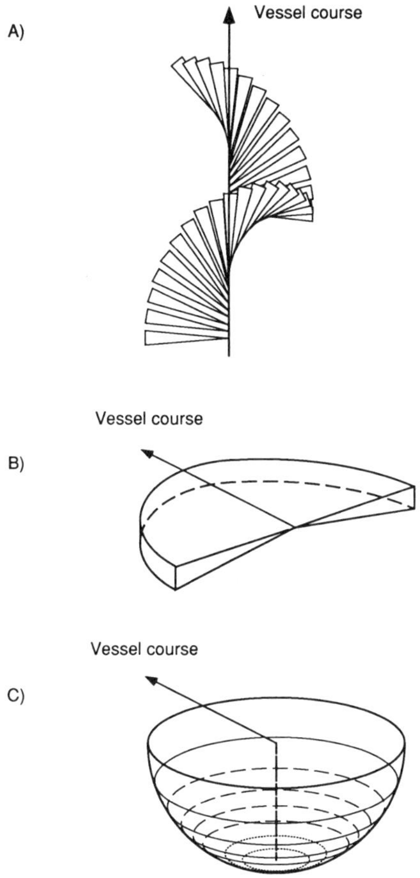

Figure 5-7. The area insonified by (A) multiple transmissions from a single beam sonar on automatic scanning mode, (B) a single transmission of a multibeam sonar scanning a 180° by 10° sector, and (C) multiple transmissions from a multibeam ("omni") sonar sequentially scanning 360° sectors.

of the 360° arc surrounding the vessel. However, examination of the scanning pattern for a single beam sonar programmed to repeatedly sweep a 180° sector in front of the survey vessel (Fig. 5-7A) clearly shows the shortcoming of these systems in quantitative surveys. Unless the vessel progresses very slowly, it is not possible to achieve complete coverage of the 180° observation field in front of the vessel. As a result, quantitative surveys with single beam systems are usually made with the sonar beam pointed perpendicular to the vessel's course (athwartships). However, it is often not possible to census the 100–200 m closest to the vessel (Hewitt et al. 1976) due to avoidance by the target species, and the small volume insonified.

Multibeam systems remedy this problem by transmitting over the entire sector (180° in Figure 5-7B) to be covered, then electronically scanning a multibeam array at a rapid rate (thousands of times per second) to achieve virtually continuous coverage of the sector. As a result, the term "within-pulse sector scanning sonar" is sometimes used to describe these systems (Mitson 1983). They are fundamentally different from mechanically steered single beam systems, despite the fact that the latter are sometimes referred to as "sector scanning" or "scanning" sonars.

Some multibeam systems are capable of covering a 360° sector with each transmission ("omnidirectional" or "omni" transmission, Fig. 5-7C). These systems can be interfaced with the navigation and trawl mensuration systems, and are capable of showing the relative location of the vessel, fishing gear (trawl or purse seine), and fish schools in a single display. Some omni systems can display vertical profiles of schools, making it possible to estimate their height and volume.

Remote echo integration of school biomass using multibeam sonar seems feasible, but problems with noise reverberation, attenuation of acoustic energy by dense schools, and variation in target strength with aspect presented to the sonar beam have yet to be overcome. At present then, the most reliable method of determining school biomass (S_i) is to capture the entire school with a purse seine. This is a difficult, time consuming process, and sample sizes will consequently be small. When school biomass (S_i) is strongly correlated with school area (a_i) or volume (v_i), it may be possible to estimate the biomass of all schools detected by measuring a_i or v_i, then estimating S_i from a relationship such as

$$\hat{S}_i = ka_i$$

with k being estimated from a sample of schools where a_i is measured with the sonar, then S_i determined by purse seining.

To achieve a strong correlation between a_i and S_i, packing density would have to vary little between schools. However, an extensive study off Norway and in the North Sea (Misund 1991) showed that packing density for herring, sprat, and saithe declined with the size of fish in the schools, and also showed substantial geographical and seasonal differences. Within geographic regions, packing density varied by as much as a factor of 100 between schools of the same species and size composition.

Errors involved in measuring school dimensions using sonar also contribute to variability in the relationship between a_i and S_i, and in the few studies made to date, correlations between a_i (as well as school volume) and S_i have proven significant in some cases (Misund 1990) and nonsignificant in others (Hewitt 1976). In general, it seems unlikely that useful estimates of mean school biomass can be attained without making a substantial number of purse seine sets. The sampling program for determining mean school biomass must parallel the program for estimating school density, and perhaps equal it in scope.

Camera Sleds and Submersibles

While strip transects are most common when deploying camera sleds, manned submersibles, and ROVs, line transect surveys are a viable alternative in some situations. Use of the line transect method requires estimates of sighting distances and angles, however, and when the target species is highly mobile or abundant, the collection of these data may not be practical.

A modified SCUBA diver's hand-held sonar can be pressed against the viewing window of a submersible to estimate distances to the bottom, prominent rocks, and cliff faces. This technique has been used to estimate perpendicular distances for yelloweye rockfish (*Sebastes ruberrimus*) off Southeast Alaska, and determine the effective sampling width using a hazard rate model (V. O'Connell, ADFG, personal communication). More sophisticated acoustic systems, capable of making direct measurements of perpendicular distances to fish targets, should become available in the near future.

The feasibility of carrying out a line transect survey using an underwater video camera mounted on a towing sled was evaluated by Bergstedt and Anderson (1990), who tested the performance of various models in estimating the density of known populations of bricks placed on the bottom of Lake Huron. Several surveys were made, with at least 40 observations being made in each survey. They found that the Fourier series estimator yielded unbiased estimates of density with CVs ranging from 15 to 20% for the different surveys. Density was overestimated by 6.3% when using the Fourier series, and underestimated by 17.6% when using the hazard rate model.

Butler et al. (1991) used the hazard rate model to estimate the effective width sampled by a camera sled using still photography. Perpendicular distance data were obtained for three species of fish using the photogrammetric techniques described in Wakefield and Smith (1990).

Complete Counts

In some surveys, the entire population is counted, and there is no need to determine the area sampled or estimate density (number/m^2). This is most common in the case of species that migrate along a narrow corridor, where the observer can count each individual as it passes. This can be likened to counting the objects on a conveyer belt as they pass the observer. Direct counts are common in the case of anadromous fish such as salmon and alewives, and a variety of dams, weirs, and counting fences have been constructed in streams over the years in order to restrict the width of observation area, and aid the counting process. Further aids include the construction of special counting towers, deployment of turbulence reducers on the surface, and placement of background panels on the streambed to enhance the visibility of the target species. Polaroid sunglasses can greatly enhance the effectiveness of the counting process.

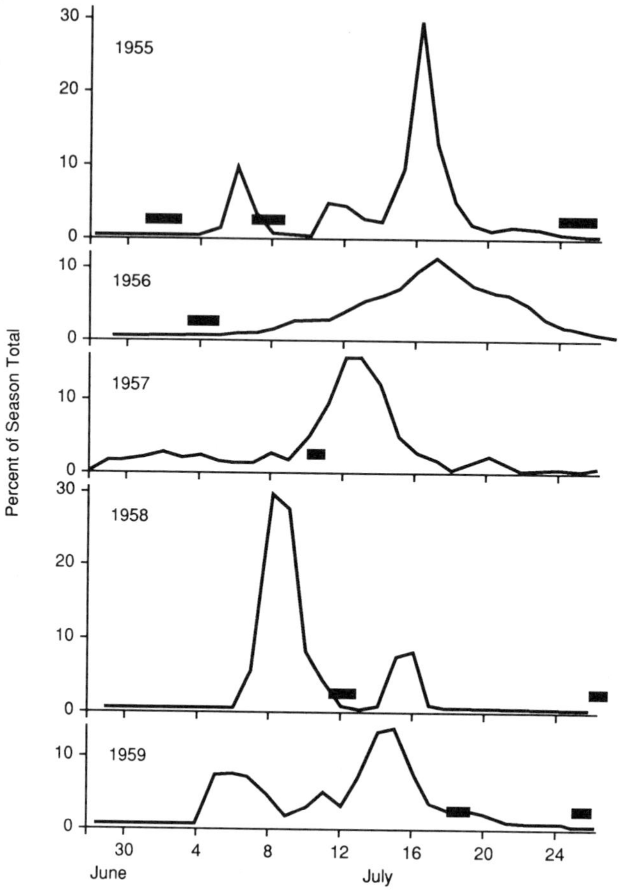

Figure 5-8. Temporal distribution of counts of migrating sockeye salmon in the Kvichak River, 1955–1959. Solid horizontal bars indicate periods of high turbidity. After Becker (1962).

Migratory mammals such as gray and bowhead whales have also been censused in this way when they pass near the shore. This is usually accomplished by using telescopes mounted on towers, or deployed from natural outlook points on the land. Optical range finders can be used to determine distances and theodolites to determine sighting angles. Landbased counts of migrating whales must often be supplemented with special aerial surveys in order to estimate the proportion of the population that migrates seaward of the visual range of the observers.

Routine counts can typically be made only during the day regardless of the target species, so that special studies must be undertaken to estimate the proportion of migrants that pass during the night. In the case of anadromous species nocturnal migrants can sometimes be trapped and held until the observer is ready to count them.

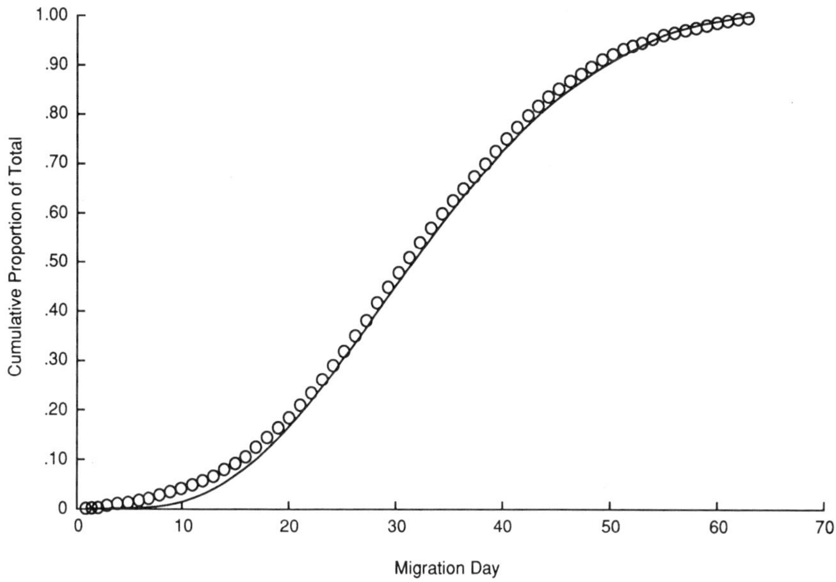

Figure 5-9. Cumulative proportion of the count of gray whales migrating past Monterey, California (circles) and the line fit to the cumulative gamma model. After Reilly et al. (1983).

Observers become fatigued if counting continues over a prolonged period, so that some form of random or systematic subsampling must be included in the design. A number of studies (Becker 1962; Jessop and Harvie 1990) have shown that the highest precision is attained with short, frequent counts (10–15 min/hr). The counts usually follow a systematic design, since these are easier to implement than randomized designs.

Migrations often follow a bell-shaped curve, with few migrants near the tails, and the bulk of the run occurring during the peak (Fig. 5-8). In the case of sockeye salmon (*Oncorhynchus nerka*) in the Kvichak River, the bulk of the migration usually passes the counting towers within a 3-week period, and it is important that sampling intensity be increased near this peak (Becker 1962). The shape of the migration curve has sometimes been modeled with a normal or gamma distribution, with the goal of developing run-time curves (Fig. 5-9) that will eliminate the need to sample throughout the entire migration period (Mundy 1979; Reilly et al. 1983). The use of these curves should be approached with caution, however, since wide interannual variations in run timing characterize many species (Fig. 5-8).

Complete counts of nonmigratory species in streams have also been undertaken, although the costs involved usually preclude sampling the entire stream. In this situation, each individual pool, riffle, or run (glide) becomes a sampling unit. A subsample of these units is selected for each habitat type, and the entire population is counted within each of them (Hankin and Reeves 1988, see Examples of Successful Applications).

EXAMPLES OF SUCCESSFUL APPLICATIONS

Tower Counts of Sockeye Salmon

The Kvichak River system of western Alaska is the foremost producer of sockeye salmon (*Oncorhynchus nerka*) for the Bristol Bay commercial fishery. Because of the importance of this fishery, there is a long history of attempts to estimate the escapement (i.e., the number of migrants that escape the fishery and migrate upstream to spawn) for management purposes. In 1932 the entire escapement was counted by installing a weir on the upper river, but this practice was discontinued due to the high costs associated with the construction and maintenance of weirs. Estimates of the numbers of salmon on the spawning grounds have also been obtained from aerial surveys, but failed to reveal more than 20% of the estimate obtained from tower counts (Becker 1962).

A system of visual counts was initiated in 1955, and was designed to take advantage of the migratory behavior of sockeye salmon (Becker 1962). Once in the river the fish seek places where currents are reduced. Since areas of low currents are usually found near the bottom and next to the banks, the migrants follow restricted paths close to the shore where they can be readily observed. The direction of migration is usually continuously upriver.

Counting towers were erected at a section of the stream where the moving bands of fish were constricted in width, and passed near the base of the structures (Fig. 5-10). Towers 1 and 2 provided for sampling the two primary paths in the main channel, which had a width of 116 m and a maximum depth of about 5 m. During sunny, calm days when the water was clear, salmon could be seen at any point in the river between towers 1 and 2. Tower 3 covered a shallow secondary channel through which fish migrated when the water level was high. Species other than sockeye were omitted from the counts when identified, although extensive beach seining has shown that salmon other than sockeye make up less than 0.5% of the total escapement.

Light gray panels were placed on the river bottom to silhouette the passing salmon. As a rule the fish pass readily over any panel of dark or dull tones in preference to panels of bright tones. Turbulence reducers constructed of boards and logs were placed immediately upstream of the panels to smooth the surface of the water and increase the visibility of the migrants. Counters used polaroid glasses during the daylight hours, and deployed spotlights at night (typically 2300–0200). The spotlights were focused outside of the panels from the top of the towers, so that the fish passed readily in the dimly lit area inshore of the spotlight, yet could be seen satisfactorily when crossing the background panels.

Sampling was systematic, with observers at each tower taking 10 minute counts each hour during most of the run. Sampling intensity was reduced (one 10 minute count every 1–4 hours) at the beginning and end of the run, and increased (15–20 minute counts each hour) during peak migration. In-

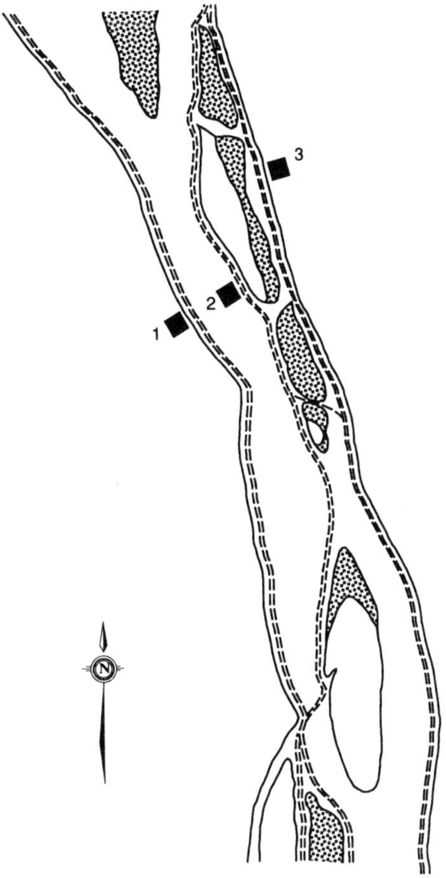

Figure 5-10. Location of counting towers (numbered squares) and paths of migrating fish (dashed lines) on the Kvichak River. The unstippled areas between the river banks are islands; submerged bars or shallows are denoted by stippling. After Becker (1962).

dividual fish were counted when fewer than 250 fish passed per minute, but tallied by estimating groups of 10 when this threshold was exceeded. Whenever systematic counts ceased at night or when the waters were turbid with sediment, the number of fish passing was estimated by averaging the counts preceding and following the period with missed counts. Hourly estimates of escapement were obtained by extrapolating the counts for each sampling interval to the entire hour, and daily estimates by summing the hourly estimates from each tower.

Controlled experiments with paired observers showed that individual counters differed by no more than about 10.7%, with an average difference of 1%. Interruptions in counting that result from periods of high turbidity could potentially result in serious bias if they occur during peaks in the run, but this did not appear to occur during 1955–1959 (Fig. 5-8). Beach seine hauls indicate that the salmon tend to avoid turbid water, so migration during periods of turbid water were based on an average of low counts immediately preceding and following the period of turbidity.

Kvichak River escapements were estimated to be 250,500 fish in 1955, 9,443,300 in 1956, 2,842,800 in 1957, 534,800 in 1958, and 680,000 in 1959. The variance of the 1959 estimate was estimated by poststratifying the data into four strata on the basis of mean daily counts, and treating the counts in each stratum as random variables. The means for each stratum were then weighted by the proportion of the observations that occurred within that stratum, and a variance estimate obtained for the sum of these weighted means. The 95% confidence limits for the 1959 escapement estimate were ±3.99%, although this probably overstates the precision of the estimate due to the poststratification procedure employed.

Visual tower counts for sockeye salmon are currently used widely throughout the Bristol Bay system, providing escapement estimates that serve as the basis for managing these fisheries. Managers typically have total escapement goals for each river, and the fishery is opened and closed throughout the period of migration to ensure that these goals are reached. The feedback provided by the tower counts is consequently a central element in the proper management of these fisheries.

Snorkeling Counts in Streams

Estimation of population sizes in streams presents significant sampling problems when the entire stream must be censused. Since it is seldom feasible to survey the entire stream, two-stage sampling designs are common. A random or systematic selection of sampling units is selected during the first stage, and the population of fish within each of these is estimated during the second. Hankin (1984) concluded that errors of estimation of fish numbers within selected units (second stage errors) are likely to be small compared with errors that arise due to variation in fish numbers between these units (first stage errors).

Hankin and Reeves (1988) used natural habitat units as the basic sampling element in designing snorkeling surveys of salmonids in Cummins Creek, Oregon, classifying each habitat unit as either a riffle, pool, glide (run), or side channel. The stream was further stratified on the basis of location (e.g., upper, middle, or lower pools). The length and area encompassed by each habitat unit varied.

The habitat units to be censused (first stage sample) were selected systematically, every fifth unit being sampled (using independent random starts for each habitat type/location stratum). One team classified each habitat unit, estimated its area, and marked those units in which fish numbers were to be counted by divers 2–5 hours later. Stream width ranged from 2 to 16 m and mean depth of individual habitat units ranged from 0.1 to 1.3 m. Side channels and glides constituted less than 20% of the total number of identified units and less than 5% of the estimated total area of all habitat unit types.

Independent counts of fish were made by a team of two divers, using mask and snorkel, for each unit that appeared in the systematic sample. Divers entered the water downstream from a selected unit and proceeded slowly upstream, identifying species and/or age class and counting individual fish. The two divers positioned themselves near the midline of a selected unit and moved through it parallel to one another, using hand signals to coordinate movements. All counts were made between 0900 and 1600 when visibility was good, averaging 4–5 m. Snorkeling was judged to be an effective technique for counting juvenile coho salmon (*Oncorhynchus kisutch*) and 1+ steelhead trout (*O. mykiss*), but noneffective for 0+ steelhead trout. Agreement between diver counts was good for most sample units.

Diver counts were calibrated for each fish species and habitat type against a more accurate method of estimation by selecting a systematic subsample (one out of every three or four) of the units that had been censused. Successive electrofishing passes were made through each calibration unit, and the Moran–Zippen removal method (Seber 1973) used to obtain a population estimate that was independent of the diver counts. Electrofishing passes were continued in a given unit until the number of fish captured was less than 10% of the number removed during the previous pass. Estimated ratios of actual fish present to mean diver counts (R) were 1.01 and 1.36 for coho salmon in pools and riffles, respectively, and 0.97 and 1.05 for steelhead trout. Bias could still exist, however, since experiments with a number of species have shown that depletion experiments involving repeated electrofishing can seriously underestimate the true population (Cross and Stott 1975; Peterson and Cederholm 1984). This is due to a decline in catchability that occurs when short time periods (less than 24 hours) separate successive removals.

The number of fish in habitat unit i was estimated by

$$\hat{P}_i = d_i R$$

where $\hat{P}_i$ = estimated population size in habitat unit i
d_i = mean diver count in habitat unit i
R = $\Sigma_i \hat{P}_{ei} / \Sigma_i d_i$, where $\hat{P}_{ei}$ is the population estimate obtained in selected units by electrofishing

An analysis of the data showed that the area encompassed by each habitat unit was poorly correlated with the number of fish present, so the total number of fish within a habitat type/location stratum ($\hat{P}_h$) was estimated using the formula for a systematic sample. A more complex estimator (e.g., two-stage ratio estimator), incorporating a measure of the size of each primary unit selected, may be preferable to the one presented here if the population size in the habitat units is highly correlated with area (Hankin 1984). The estimator used in Cummins Creek was

$$\hat{P}_h = \frac{N_h}{n_h} \sum_{i=1}^{n_h} \hat{P}_{hi}$$

where $\hat{P}_{hi}$ = estimated population size for the ith sampling unit in habitat type h

N_h = total number of habitat units of type h present in the survey area

n_h = number of habitat units of type h in the systematic sample

The variance of this estimate is a function of the variance of R, between-diver variance, and the variance in population size between habitat units. An estimator for Var($\hat{P}$), which takes all of these sources of variance into account, was developed by Hankin and Reeves (1988). Estimated population sizes in all pools and riffles (the two principal habitat types) were 4106 coho salmon and 3409 1+ steelhead trout, with 95% confidence limits of ±22 and ±17%, respectively.

The total number of habitat units for which fish numbers could be estimated was dramatically increased through the use of visual snorkeling counts. This increased first stage sampling fraction resulted in a substantial reduction in total errors of estimation, since the variance contributions associated with the second stage of sampling (within habitat units) were relatively small.

Correlations between diver counts and removal estimates of population size were quite high ($r \geq 0.94$) except for steelhead trout in pools ($r = 0.61$). Steelhead were closely associated with the bottom substrate, often difficult to see, and moved to cover more quickly than coho. In contrast, coho salmon were more surface oriented, and easier to observe.

The wisdom of stratification by habitat type and location was supported by strong differences in average fish densities between these strata. Mean estimated densities of fish per unit area were much higher in pools and riffles from the lower and middle reaches than from the upper reach. Densities of fish per unit area in pools were 5–10 times those in riffles within a given location. These examples suggest that extrapolation from data collected in "representative" reaches or habitat types could give a highly biased or misleading picture of true fish abundance.

Camera Sled Surveys

Deep-water surveys of the demersal fish fauna off Monterey Bay, California have been carried out using both bottom trawls and camera sleds. The longspine thornyhead (*Sebastes altivelis*) is a dominant member of the assemblage of upper slope fishes, and has been the subject of considerable study to determine its importance in the vertical transport of particulate organic matter over the course of its life history (Wakefield and Smith 1990).

The apparently passive behavior of thornyhead in response to manned submersibles and camera sleds makes this species a good candidate for photographic surveys. No evidence of avoidance or attraction has ever been observed from underwater vehicles themselves, although more thorough evaluation of these phenomena would probably be desirable, using methodology capable of monitoring behavior well in advance of the camera sled.

The abundance of thornyheads was estimated from counts of individuals in photographs taken with a camera sled towed along bottom transects (Wakefield 1990). A total of 30 camera sled transects, stratified along isobaths (200 m intervals), yielded 21,493 photographs, 87% of which were judged to be of adequate quality for estimating density. The camera sled was constructed of a welded steel frame weighing 517 kg in air.

The camera system consisted of two forward looking 35-mm cameras mounted in stereo configuration (40 cm separation) with the lens nodal points 81 cm above the the sled's skids (76 cm above the sea floor, assuming the skids sank 5 cm into the sediment). The inclination of the optical axis (16° below the horizon line, Figure 5-3), and a 50-mm, water-corrected lens resulted in an oblique view, in which the lower edge of the picture frame represented a line on the sea floor 1.2 m in front of the camera sled. Illumination was provided by a 100 J/sec strobe light. The camera was controlled by a timing circuit that included an adjustable delay and exposure interval.

The camera sled was deployed on a trawl wire with a 150-kg depressor weight attached 60–80 m above the sled. A 12-kHz pinger (transmitter) was attached to the wire 5 m above the depressor weight, and the difference in arrival times between the direct and bottom-reflected signal used to locate the sea floor relative to the weight. During the tow the weight was maintained at an altitude of 25–35 m, thereby preventing the towing wire from hitting the sea floor and eliciting an avoidance response. The sled was towed at 0.6 to 1.3 m/sec, and a 5.5 second interval between strobe firings resulted in a distance of 3.3–8.4 m between centers of consecutive photographic frames. Transect length depended on the length of the film used, towing speed, and bottom topography, ranging from 0.6 to 6.2 km.

Each film image was projected onto an electronic digitizer interfaced with a computer. A computer program was developed to analyze the images using a perspective grid (Fig. 5-3) to measure distances and fish lengths. Topographic changes can alter the camera's height and orientation, thus changing the quadrat area and introducing errors in distance measurements. Frames showing distortion due to changes in the sled's attitude were rejected, and errors of this nature were further minimized by confining counts to roughly the lower half of the frame.

Counting area (quadrat size) within photographs was determined a posteriori for each transect. Plots of abundance versus distance from the photograph's center line were constructed, and counting areas determined as those regions where density was similar. Counting areas ranged from 2.3 to 3.1 m^2, but were limited to the foreground (1.2 m^2) for newly recruited juveniles. Butler et al. (1991) fit the data on sighting distances to a hazard rate model (Fig. 5-5), and found that the probability of detection was relatively constant out to a perpendicular distance of 70 cm and out to 180 cm in front of the sled. This corresponds to an average quadrat size of 2.5 m^2 in the case of thornyheads.

Counts for each 10-frame interval were divided by the area of the sea floor contained within those frames to estimate the density for the interval (about

40–50 m along the transect). The mean density for each transect was then estimated from the mean of the interval estimates. Seven surveys were completed, and depth-stratified population estimates (and their variances) obtained by treating the mean of the transect density observations for each depth and survey as an independent replicate. Since the camera sled could not be deployed on hard substrates, density estimates pertained only to soft-bottom habitats.

Longspine and shortspine (*S. alascanus*) thornyheads could not be distinguished in photographs, so the proportion of these species in trawl catches was used to apportion the density estimates between species. A semiballoon trawl was used, and had a 15.2-m footrope rigged with 0.5-cm steel chain, hung in loops. A single towing warp was used, and the trawl doors were linked to the towing warp by a 12.5-m bridle, and to the net by a 1.9-m bridle. On-bottom duration of each trawl haul was measured by mounting a time–depth recorder on one of the trawl doors. Mouth opening and door spread were not monitored during the survey, but the mouth opening was estimated to be 5.7 m based on measurements made for a trawl with similar dimensions. During the survey period, 36 bottom trawl hauls were completed between 400 and 1300 m.

Population estimates based on photographic analysis showed that 83% of the population of longspine thornyheads reside within an area of the slope bounded by the 500- and 1100-m isobaths (Wakefield and Smith 1990). Longspine thornyheads dominated trawl catches of thornyheads in this depth interval, increasing from 97% of the catch at 400 m to 99% at 1000 m. The total population of thornyheads in a representative 100-m-wide strip between the 500- and 1100-m isobaths was estimated to be 107,000 fish, with approximate (2 SE) 95% confidence limits of ±19%.

Estimates of density (weight per square meter) based on camera transects were consistently greater than those obtained using the area swept method for the trawl data. The differences ranged from a factor of 1.3 in the 800-m stratum to a factor of 4.9 in the 1200-m stratum, with an average of 2.7 over all strata. These results clearly indicated that the trawl performed poorly in terms of quantitative estimation of population density, although the exact mechanisms involved could not be determined without more complete mensuration of trawl performance (wing spread, door spread, footrope height off bottom, etc.) and examination of the behavior of the fish in response to the gear and towing bridles. Size composition estimates based on photogrammetric measurements were similar to those obtained in trawl samples.

BIBLIOGRAPHY

Aglen, A. 1989. Reliability of acoustic fish abundance estimates. Dr. Scient. Thesis, University of Bergen, 106 pp.

Alheit, J. 1985. Spawning frequency of Peruvian anchovies taken with a purse seine. Pp. 59–61 in R. Lasker (ed.), An Egg Production Method for Estimating Spawning Biomass of Pelagic Fish: Application to the Northern Anchovy, *Engraulis mordax*. NOAA Technical Report NMFS 36.

Alheit, J. 1988. Reproductive biology of sprat (*Sprattus sprattus*): Factors determining annual egg production. J. Cons. Int. Explor. Mer 44:162–168.

Alheit, J. 1989. Comparative spawning biology of anchovies, sardines, and sprats. Rapp. P.-v. Reun. Cons. Int. Explor. Mer 191:7–14.

Alheit, J. 1993. Application of daily egg production method to clupeioid fish. In J.R. Hunter and N.C.H. Lo (eds.), Ichthyoplankton Methods for Estimating Fish Biomass. Sci. Ser. Los Angeles County Nat. Hist. Museum (in press).

Alheit, J., V.H. Alarcon, and B.J. Macewicz. 1984. Spawning frequency and sex ratio in the Peruvian anchovy, *Engraulis ringens*. Calif. Coop. Oceanic Fish. Invest. Rep. 25:43–52.

Angelson, K.K., and S. Olsen. 1987. Impact of fish density and effort level on catching efficiency of fishing gear. Fish. Res. 5:271–278.

Armstrong, M., P. Shelton, I. Hampton, G. Jolly, and Y. Melo. 1988. Egg production estimates of anchovy biomass in the Southern Benguela system. Cal. Coop. Fish. Invest. Rep. 29:137–157.

Atkinson, D.B. 1989. Diel movements of beaked redfish and the implications of these for stratified random bottom trawl estimates of biomass and abundance. N. Am J. Fish. Mgt. 9:163–170.

Azarovitz, T.R. 1981. A brief historical review of the Woods Hole Laboratory trawl survey time series. Pp. 62–67 in W.G. Doubleday and D. Rivard (eds.), Bottom Trawl Surveys. Can. Spec. Publ. Fish. Aquat. Sci. 58, 273 pp.

Bakkala, R.G., and T.K. Wilderbuer. 1991. Yellowfin sole. Pp. 86–105 in Stock Assessment and Fishery Evaluation Document for Groundfish Resources in the Bering Sea/Aleutian Islands Region as Projected for 1992. North Pacific Fishery Mgt. Council, Unpubl. Rept., 426 pp.

Becker, C.D. 1962. Estimating red salmon escapements by sample counts from observation towers. U.S. Fish Wildlife Service Fish. Bull. 61:355–369.

Bergstedt, R.A., and D.R. Anderson. 1990. Evaluation of line transect sampling based on remotely sensed data from underwater video. Trans. Am. Fish. Soc. 119:86–91.

Berrien, P.L., N.A. Naplin, and M.R. Pennington. 1981. Atlantic mackerel, *Scomber scombrus*, egg production and spawning population estimates for 1977 in the Gulf of Maine, Georges Bank, and Middle Atlantic Bight. Rapp. P.-v. Reun. Cons. Int. Explor. Mer 178:279–288.

Biggs, E.D., and F. Funk. 1988. Pacific herring spawning ground surveys for Prince William Sound, 1988, with historic overview. Alaska Dept. Fish and Game Regional Info. Rept. 2C88–07, 27 pp. plus appendices.

Blaxter, J.H.S. 1969. Development: Eggs and larvae. Pp. 177–252 in W. S. Hoar and D.G. Randall (eds.), Fish Physiology, Vol. 3. Academic Press, New York.

Blaxter, J.H.S. 1986. Development of sense organs and behavior of teleost larvae with special reference to feeding and predator avoidance. Trans. Am. Fish. Soc. 115:98–114.

Blaxter, J.H.S., and R.S. Batty. 1990. Swimbladder "behavior" and target strength. Rapp. P.-v. Reun. Cons. Int. Explor. Mer 189:233–244.

Bolin, R.L. 1936. Embryonic and early larval stages of the California anchovy *Engraulis mordax* Girard. Calif. Fish Game 22:314–321.

Bradford, M.J., and R.M. Peterman. 1989. Incorrect parameter values used in virtual population analysis (VPA) generate spurious time trends in reconstructed abundances. Can. Spec. Publ. Fish. Aquat. Sci. 108:87–99.

Brander, K., and A. B. Thompson. 1989. Diel differences in avoidance of three vertical profile sampling gears by herring larvae. J. Plankton Res. 11:775–784.

Bridger, J.P. 1969. The behavior of demersal fish in the path of a trawl. FAO Fish. Rep. 62:695–716.

Brock, R.E. 1982. A critique of the visual census method for assessing coral reef fish populations. Bull. Mar. Sci. 32:269–276.

Buchanan-Wollaston, H.J. 1926. Plaice-egg production in 1920–21, treated as a statistical problem, with comparison between the data from 1911, 1914, and 1921. Fish. Invest. Ser. II, 9(2). 45 pp.

Buckland, S.T. 1985. Perpendicular distance models for line transect sampling. Biometrics 41:177–195.

Buerkle, U. 1977. Detection of trawling noise by Atlantic cod (*Gadus morhua* L.). Mar. Behav. Physiol. 4:233–242.

Buerkle, U., and R.L. Stephenson. 1991. Herring school dynamics and its impact on acoustic abundance estimates. Pp. 185–207 in Proc. Int. Herring Symposium Anchorage, Alaska, October 23–25, 1990. Alaska Sea Grant Rep. 91–01.

Burczynski, J.J. 1979. Introduction to the use of sonar systems for estimating fish biomass. FAO Fisheries Technical Paper No. 191, 89 pp.

Burczynski, J.J., and R.L. Johnson. 1986. Application of dual-beam acoustic survey techniques to limnetic populations of juvenile sockeye salmon (*Oncorhynchus nerka*). Can. J. Fish. Aquat. Sci. 43:1776–1788.

Burnham, K.P., D.R. Anderson, and J.L. Laake. 1980. Estimation of density from line transect sampling of biological populations. Wildlife Monographs No. 72. 202 pp.

Butler, J.L., W.W. Wakefield, P.B. Adams, B.H. Robinson, and C.H. Baxter. 1991. Application of line transect methods to surveying demersal communities with ROVs and manned submersibles. Proc. IEEE Oceans '91 Conf, pp. 689–696.

Byrne, C.J., T.R. Azarovitz, and M.P. Sissenwine. 1981. Factors affecting variability of research vessel trawl surveys. Pp. 258–273 in W.G. Doubleday and D. Rivard (eds.), Bottom Trawl Surveys. Can. Spec. Publ. Fish. Aquat. Sci. 58, 273 pp.

Carlson, H.R., and R.R. Straty. 1981. Habitat and nursery grounds of Pacific rockfish, *Sebastes* spp., in rocky coastal areas of southeastern Alaska. Mar. Fish. Rev. 43:13–19.

Carney, R.S., and A.G. Carey, Jr. 1980. Effectiveness of metering wheels for measurement of area sampled by beam trawls. Fish. Bull. 3:791–796.

Clark, S.H. 1981. Use of trawl survey data in assessments. Pp. 82–92 in W.G. Doubleday and D. Rivard (eds.), Bottom Trawl Surveys. Can. Spec. Publ. Fish. Aquat. Sci. 58, 273 pp.

Cochran, W.G. 1977. Sampling Techniques, 3rd ed. John Wiley, New York, 428 pp.

Collie, J. S., and M. P. Sissenwine. 1983. Estimating population size from relative abundance data measured with error. Can. J. Fish. Aquat. Sci. 40:1871–1879.

Cram, D.L., and I. Hampton. 1976. A proposed aerial/acoustic strategy for pelagic fish stock assessment. J. Cons. Int. Explor. Mer 37:91–97.

Crecco, V., and W.J. Overholtz. 1990. Causes of density-dependent catchability for Georges Bank haddock *Melanogrammus aeglefinus*. Can. J. Fish. Aquat. Sci. 47:385–394.

Creutzberg, F., G.C.A. Duineveld, and G.J. van Noort. 1987. The effect of different numbers of tickler chains on beam trawl catches. J. Cons. Int. Explor. Mer 43:159–168.

Crittenden, R.N., and G.L. Thomas. 1987. Estimation of fish swimming speed and its effect on hydroacoustic measurement of fish abundance. Paper 024 presented at the International Symposium of Fisheries Acoustics, June 22–26, 1987, Seattle, WA.

Crittenden, R.N., G.L. Thomas, D.A. Marino, and R.E. Thorne. 1988. A weighted duration-in-beam estimator for the volume sampled by a quantitative echo-sounder. Can. J. Fish. Aquat. Sci. 45:1249–1256.

Cross, D.G., and B. Stott. 1975. The effect of electric fishing on the subsequent capture of fish. J. Fish. Biol. 7:349–357.

Csirke, J. 1988. Small shoaling pelagic fish stocks. Pp. 271–302 in J.A Gulland (ed.), Fish Population Dynamics. John Wiley, New York, 422 pp.

Cushing, D.H. 1982. Climate and Fisheries. Academic Press, New York, 373 pp.

Dawson, J.J., and W.A. Karp. 1990. *In situ* measures of target-strength variability of individual fish. Rapp. P.-v. Reun. Cons. Int. Explor. Mer 189:264–273.

Dixon, W.J., M.B. Brown, L. Engleman, M.A. Hill, and R.I. Jennrich. 1988. BMDP Statistical Software Manual, Vol. 2. Univ. of Cal. Press, Berkely, 1234 pp.

Doubleday, W.G., and D. Rivard (eds.). 1981. Bottom Trawl Surveys. Can. Spec. Publ. Fish. Aquat. Sci. 58, 273 pp.

Downing, J.A. 1979. Aggregation, transformation, and the design of benthos sampling programs. J. Fish. Res. Board Can. 36:1454–1463.

Draper, N.R., and H. Smith. 1981. Applied Regression Analysis, 2nd ed. John Wiley, New York, 709 pp.

Edwards, R., and J.H. Steele. 1968. The ecology of 0-group plaice and common dabs at Loch Ewe. I. Population and food. J. Exp. Mar. Biol. Ecol. 2:215–238.

Efron, B. 1982. The jackknife, the bootstrap, and other resampling plans. SIAM, CBMS-NSF Regional Conference Series in Applied Mathematics, No. 38, Philadelphia, PA.

Efron, B., and G. Gong. 1983. A leisurely look at the bootstrap, the jackknife, and cross-validation. Am. Statist. 37:36–48.

Eggers, D.M., N.A. Rickard, D.G. Chapman, and R.R. Whitney. 1982. A methodology for estimating area fished for baited hooks and traps along a ground line. Can. J. Fish. Aquat. Sci. 39:448–453.

Ehrenberg, J.E. 1979. Calculation of the constants needed to scale the output of an echo integrator. BIOMASS Handbook 7. SCAR/SCOR/IABO/ACMRR Group of Specialists on Living Resources of the Southern Oceans, 11 pp.

Ehrenberg, J.E. 1983. A review of *in situ* target strength estimation techniques. FAO Fish. Rep. 300:85–90.

Elliott, J.M. 1971. Some methods for the statistical analysis of samples of benthic invertebrates. Freshwater Biol. Assoc. Sci. Publ. 25, 148 pp.

Ellis, I.E. 1973. A progress report on the development of a selective shrimp trawl. FAO Fish. Rep. 139:34–49.

Emerson, L.S., M. Greer Walker, and P.R. Witthames. 1990. A stereological method for estimating fish fecundity. J. Fish Biol. 36:721–730.

Engås, A., and O.R. Godø. 1989a. The effect of different sweep lengths on the length composition of bottom-sampling trawl catches. J. Cons. Int. Explor. Mer 45:263–268.

Engås, A., and O.R. Godø. 1989b. Escape of fish under the fishing line of a Norwegian sampling trawl and its influence on survey results. J. Cons. Int. Explor. Mer 45:269–276.

Engås, A., and E. Ona. 1990. Day and night fish distribution pattern in the net mouth area of the Norwegian bottom-sampling trawl. Rapp. P.-v. Reun. Cons. Int. Explor. Mer 189:123–127.

Engås, A., and E. Ona. 1991. A method to reduce survey bottom trawl variability. Int. Coun. Explor. Sea. CM 1991/B:39. 6 pp. (mimeo).

Engås, A., and A.V. Soldal. 1992. Diurnal variations in bottom trawl catch rates of cod and haddock and their influence on abundance indices. ICES J. Mar. Sci. 49:89–95.

Engås, A., and C.W. West. 1987. Trawl performance during the Barents Sea cod and haddock survey: Potential sources of gear-related sampling bias. Fish. Res. 5:279–286.

Erickson, A.W., L.J. Bledsoe, and M.B. Hanson. 1989. Bootstrap correction for diurnal activity cycle in census data for Antarctic seals. Mar. Mammal Sci. 5:29–56.

Erickson, A.W., and M.B. Hanson. 1990. Continental estimates and population trends of Antarctic pack ice seals. Pp. 253–264 in K.R. Kerry and G. Hempel (eds.), Antarctic Ecosystems. Springer-Verlag, Berlin.

Fiedler, P.C. 1978. The precision of simulated transect surveys of northern anchovy, *Engraulis mordax*, school groups. Fish. Bull. 76:679–685.

Finney, D.J. 1941 On the distribution of a variate whose logarithm is normally distributed. J. R. Stat. Soc. Suppl. 7:155–161.

Foote, K.G. 1978. Analysis of empirical observations on the scattering of sound by encaged aggregations of fish. Fisk. Dir. Skr. Ser. HavUnders. 16:442–455.

Foote, K.G. 1979. On representing the length dependence of acoustic target strengths of fish. J. Fish. Res. Bd. Can. 36:1490–1496.

Foote, K.G. 1980. Importance of the swimbladder in acoustic scattering by fish: A comparison of gadoid and mackerel target strength. J. Acoust. Soc. Am. 67:2084–2089.

Foote, K.G. 1990. Correcting acoustic measurements of scatterer density for extinction. J. Acoust. Soc. Am. 88:1543–1546.

Foote, K.G. 1991. Summary of methods for determining fish target strength at ultrasonic frequencies. ICES J. Mar. Sci. 48:211–217.

Foote, K.G., and E. Ona. 1985. Swimbladder cross sections and acoustic target strengths of 13 pollack and 2 saithe. Fisk. Dir. Skr. Ser. Havunders. 18:1–57.

Foote, K.G., and J.J. Traynor. 1988. Comparison of walleye pollock target strength estimates determined from in situ measurements and calculations based on swimbladder form. J. Acoust. Soc. Am. 83:9–17.

Foote, K.G., H.P. Knudsen, G. Vestnes, D.N. MacLennan, and E.J. Simmonds. 1987. Calibration of acoustic instruments for fish density estimation: A practical guide. ICES Cooperative Research Report 144, 69 pp.

Forbes, S.T., and O. Nakken. 1972. Manual of methods for fisheries resource survey and appraisal. Part 2: The use of acoustic instruments for fish detection and abundance estimation. FAO Manuals in Fisheries Science No. 5, 138 pp.

Forest, A., and J.P. Minet. 1981. Abundance estimates of the trawlable resources around the islands of St. Pierre and Miquelon (NAFO Subdiv. 3Ps): Methods used during the French research surveys and discussion of some results. Pp. 68–81 in W.G. Doubleday and D. Rivard (eds.), Bottom Trawl Surveys. Can. Spec. Publ. Fish. Aquat. Sci. 58, 273 pp.

Francis, R.I.C.C. 1984. An adaptive strategy for stratified random trawl surveys. N.Z.J. Mar. Freshwater Res. 18:59–71.

Francis, R.I.C.C. 1985. Two acoustic surveys of pelagic fish in Hawke Bay, New Zealand, 1980. N.Z.J. Mar. Freshwater Res. 19:375–389.

Francis, R.I.C.C. 1991. Statistical properties of two-phase surveys: Comment. Can. J. Fish. Aquat. Sci. 48:1128.

Gardiner, W.R. 1984. Estimating population densities of salmonids in deep water in streams. J. Fish. Biol. 24:41–49.

Garrod, D.J. 1977. The North Atlantic cod. Pp. 216–242 in J.A. Gulland (ed.), Fish Population Dynamics. John Wiley, New York, 372 pp.

Gavaris, S., and S.J. Smith. 1987. Effect of allocation and stratification strategies on precision of survey abundance estimates for Atlantic cod (*Gadus morhua*) on the Eastern Scotian shelf. J. Northw. Atl. Fish. Sci. 7:137–144.

Gehringer, J.W., and W. Aron. 1968. Field techniques. Pp. 87–144 in D.J. Tranter and J.H. Fraser (eds.), Zooplankton Sampling. UNESCO Monographs on Oceanographic Methodology 2. Paris.

Glass, C.W., and C.S. Wardle. 1989. Comparison of the reactions of fish to trawl gear, at high and low light intensities. Fish. Res. 7:249–266.

Glass, C.W., C.S. Wardle, and W.R. Mojsiewicz. 1986. A light intensity threshold for schooling in the Atlantic mackerel, *Scomber scombrus*. J. Fish Biol. 29(Suppl. A):71–81.

Godø, O.R., and A. Engås. 1989. Swept area variation with depth and its influence on abundance indices of groundfish from trawl surveys. J. Northw. Atl. Fish. Sci. 9:133–139.

Godø, O.R., M. Pennington, and J.H. Vølstad. 1990. Effect of tow duration on length composition of trawl catches. Fish. Res. 9:165–179.

Goodman, L.A. 1960. On the exact variance of products. J. Am Stat. Assoc. 55:708–713.

Greene, L.E., and W.S. Alevizon. 1989. Comparative accuracies of visual assessment methods for coral reef fishes. Bull. Mar. Sci. 44:899–912.

Gulland, J.A. 1983. Fish Stock Assessment: A Manual of Basic Methods. John Wiley, New York, 222 pp.

Gunderson, D.R. 1976a. Results of comparative trawling experiments conducted in the eastern Bering Sea. Pp. 33–50 in Demersal Fish and Shellfish Resources of the Eastern Bering Sea in the Baseline Year 1975, Data Appendices. Northwest Fisheries Center Processed Report, 534 pp.

Gunderson, D.R. 1976b. Studies of sampling techniques conducted during the rockfish pilot survey. Unpubl. Rept., Nat. Mar. Fish. Service, 8 pp.

Gunderson, D.R., and I.E. Ellis. 1986. Development of a plumb staff beam trawl for sampling demersal fauna. Fish. Res. 4:35–41.

Gunderson, D.R., and T.M. Sample. 1980. Distribution and abundance of rockfish off Washington, Oregon, and California during 1977. Mar. Fish. Rev. 42(3–4)2–16.

Gunderson, D.R., G.L. Thomas, P. Cullenberg, and R.E. Thorne. 1981. Rockfish investigations off the coast of Washington and Oregon. Fish. Res. Inst.-Univ. Wash. Rept. 8125, 45 pp.

Gunderson, D.R., D.A. Armstrong, Y.-B. Shi, and R.A. McConnaughey. 1990. Patterns of estuarine use by juvenile English sole (*Parophrys vetulus*) and Dungeness crab (*Cancer magister*). Estuaries 13:59–71.

Hankin, D.G. 1984. Multistage sampling designs in fisheries research: Applications in small streams. Can. J. Fish. Aquat. Sci. 41:1575–1591.

Hankin, D.G., and G.H. Reeves. 1988. Estimating total fish abundance and total habitat area in small streams based on visual estimation methods. Can. J. Fish. Aquat. Sci. 45:834–844.

Hara, I. 1989. Errors in the stock assessment of the Japanese sardine (*Sardinops melanosticta*) using the line-transect method. Can. Special Publ. Fish. Aquat. Sci. 108:43–50.

Harden-Jones, F.R. 1968. Fish Migration. Edward Arnold, London, 325 pp.

Harden-Jones, F.R., A.R. Margetts, M. Greer Walker, and G.P. Arnold. 1977. The efficiency of the Granton otter trawl determined by sector-scanning sonar and acoustic transponding tags. Rapp. P.-v. Reun. Cons. Int. Explor. Mer 170:45–51.

Harling, W.R., and D. Davenport. 1977. G.B. Reed groundfish cruise No. 77-3, August 22 to September 8, 1977. Fisheries and Environment Canada, Fisheries and Mar. Service Data Rept. 42, 46 pp.

Hauser, J. W., and M. P. Sissenwine. 1991. The uncertainty in estimates of the production of larval fish derived from samples of larval abundance. ICES J. Mar. Sci. 48:23–32.

Hawkins, A.D. 1986. Underwater sound and fish behavior. Pp. 114–151 in T.J. Pitcher (ed.), The Behavior of Teleost Fishes. Johns Hopkins Univ. Press, Baltimore, MD.

Hayes, D.B., and N. Buxton. 1992. Assessment of the Georges Bank haddock stock. Pap. of the Northeast Regional Stock Assessment Workshops. Res. Doc. SAW13/1, App. to CRD-92–02, Northeast Fisheries Sci. Center, Woods Hole, MA.

Hayes, R.J., and S.T. Buckland. 1983. Radial-distance models for the line-transect method. Biometrics 39:29–42.

He, P. 1991. Swimming endurance of the Atlantic cod, *Gadus morhua*, L., at low temperatures. Fish. Res. 12:65–73.

Heath, M. 1993. An evaluation and review of the ICES herring larval surveys in the North Sea and adjacent waters. In J.R. Hunter and N.C.H. Lo (eds.), Ichthyoplankton Methods for Estimating Fish Biomass. Sci. Ser. Los Angeles County Nat. Hist. Museum (in press).

Heath, M., and J. Dunn. 1990. Avoidance of a midwater frame trawl by herring larvae. J. Cons. Int. Explor. Mer 47:140–147.

Helfman, G.S. 1983. Underwater methods. Pp. 349–369 in L.A. Nielsen and D.L. Johnson (eds.), Fisheries Techniques. Am. Fish. Soc., Bethesda, MD.

Hensen, V., and C. Apstein. 1897. Die Nordsee-Expedition 1895 des Deutschen Seefischereivereins. Uber die Eimenge der im Winter laichenden Fische. Wiss. Meresunters. Helgoland 2(2), 101 pp.

Hewitt, R.P. 1985. Comparison between egg production method and larval census method for fish biomass assessment. Pl. 95–99 in R. Lasker (ed.), An Egg Production Method for Estimating Spawning Biomass of Pelagic Fish: Application to the Northern Anchovy, *Engroulis Mordax*. NOAA Technical Report NMFS 36.

Hewitt, R.P., P.E. Smith, and J.C. Brown. 1976. Development and use of sonar mapping for pelagic stock assessment in the California current area. Fish. Bull. 74:281–300.

Hiby, A.R., and P.S. Hammond. 1989. Survey techniques for estimating abundance of cetaceans. Rep. Int. Whal. Comm. Spec. issue 11:47–80.

Hiby, L., D. Thompson, and A. Ward. 1987. Improving census by aerial photography—an inexpensive system based on nonspecialist equipment. Wildl. Soc. Bull. 15:438–443.

Hilborn, R., and C.J. Walters. 1992. Quantitative Fisheries Stock Assessment: Choice, Dynamics and Uncertainty. Chapman and Hall, New York, 570 pp.

Hoff, R.Z. 1989. Spatial distributions and abundance trends for two species of flatfish in the eastern Bering Sea. M.S. thesis, University of Washington, 154 pp.

Holliday, D.V. 1977a. The use of swimbladder resonance in the sizing of schooled pelagic fish. Rapp. P.-v. Reun. Cons. Int. Explor. Mer 170:130–135.

Holliday, D.V. 1977b. Two applications of the Doppler effect in the study of fish schools. Rapp. P.-v. Reun. Cons. Int. Explor. Mer 170:21–30.

Holt, R.S., and S.N. Sexton. 1990. Monitoring trends in dolphin abundance in the Eastern Tropical Pacific using research vessels over a long sampling period: Analyses of 1986 data, the first year. Fish. Bull. 88:105–111.

Hoopes, D.T., and J. Greenough. 1968. King crab research. Int. North Pac. Fish. Comm. Annu. Rep. 1968, pp. 116–124.

Houde, E.D., P.B. Ortner, L. Lubbers, and S.R. Cummings. 1989. Test of a camera-net system to determine abundance and heterogeneity in anchovy egg distributions. Rapp. P.-v. Reun. Cons. Int. Explor. Mer 191:112–118.

Hughes, S.E. 1976. System for sampling large trawl catches of research vessels. J. Fish. Res. Board Can. 33:833–839.

Hunter, J.R. 1985. Preservation of northern anchovy in formaldehyde solution. Pp. 63–65 in R. Lasker (ed.), An Egg Production Method for Estimating Spawning Biomass of Pelagic Fish: Application to the Northern Anchovy, *Engraulis mordax*. NOAA Technical Report NMFS 36.

Hunter, J.R., and S.R. Goldberg. 1980. Spawning incidence and batch fecundity in northern anchovy, *Engraulis mordax*. Fish. Bull. 77:641–652.

Hunter, J.R., and R. Leong. 1981. The spawning energetics of female northern anchovy, *Engraulis mordax*. Fish. Bull. 79:215–517.

Hunter, J.R., and B.J. Macewicz. 1980. Sexual maturity, batch fecundity, spawning frequency, and temporal pattern of spawning for the northern anchovy, *Engraulis mordax*, during the 1979 spawning season. Cal. Coop. Fish. Invest. Rep. 21:139–149.

Hunter, J.R., and B.J. Macewicz. 1985a. Rates of atresia in the ovary of captive and wild northern anchovy, *Engraulis mordax*. Fish. Bull. 83:119–136.

Hunter, J.R., and B.J. Macewicz. 1985b. Measurement of spawning frequency in multiple spawning fish. Pp. 79–94 in R. Lasker (ed.), An Egg Production Method for Estimating Spawning Biomass of Pelagic Fish: Application to the Northern Anchovy, *Engraulis mordax*. NOAA Technical Report NMFS 36.

Hunter, J.R., N.C.H. Lo, and R.J.H. Leong. 1985. Batch fecundity in multiple spawning fishes. Pp. 67–77 in R. Lasker (ed.), An Egg Production Method for Estimating Spawning Biomass of Pelagic Fish: Application to the Northern Anchovy, *Engraulis mordax*. NOAA Technical Report NMFS 36.

Hunter, J.R., B.J. Macewicz, and J.R. Sibert. 1986. The spawning frequency of skipjack tuna, *Katsuwonus pelamis*, from the South Pacific. Fish. Bull. 84:895–903.

Hunter, J.R., B.J. Macewicz, N.C.H. Lo, and C.A. Kimbrell. 1992. Fecundity, spawning and maturity of female Dover sole *Microstomus pacificus*, with an evaluation of assumptions and precision. Fish. Bull. 90:101–128.

Hurlbert, S.H. 1984. Pseudoreplication and the design of ecological field experiments. Ecological Monographs 54:187–211.

Iribarne, O.O., M.I. Lasta, H.C. Vacas, A.M. Parma, and M.S. Pascual. 1991. Assessment of abundance, gear efficiency and disturbance in a scallop dredge fishery: results of a depletion experiment. Pp. 242–248 in S. E. Shumway and P.A. Sandifer (eds.), Scallop Biology and Culture. World Aquaculture Workshops, No. 1.

Jahn, A.E. 1987. Precision of estimates of abundance of coastal fish larvae. Am. Fish. Soc. Symp. 2:30–38.

Jessen, R.J. 1978. Statistical Survey Techniques. John Wiley, New York, 520 pp.

Jessop, B.M., and C.J. Harvie. 1990. Evaluation of designs of periodic count surveys for the estimation of escapement at a fishway. N. Am. J. Fish. Mgt. 10: 39–45.

Johannesson, K.A., and R.B. Mitson. 1983. Fisheries acoustics: A practical manual for aquatic biomass estimation. FAO Fisheries Technical Paper No. 240, 249 pp.

Joll, L. M., and J.W. Penn. 1990. The application of high-resolution navigation systems to Leslie-De Lury depletion experiments for the measurement of trawl efficiency under open-sea conditions. Fish. Res. 9:41–55.

Jolly, G.M., and I. Hampton. 1990. A stratified random transect design for acoustic surveys of fish stocks. Can. J. Fish. Aquat. Sci. 47:1282–1291.

Karp, W.A. 1990. Results of echo integration midwater-trawl surveys for walleye pollock in the Gulf of Alaska in 1990. App. 3 in Stock Assessment and Fishery Evaluation Report for the 1991 Gulf of Alaska Groundfish Fishery. North Pacific Fishery Mgt. Council, Unpubl. Rept., 293 pp.

Kautsky, G.A. 1989. *In situ* target strength estimation for the Puget Sound stock of whiting (*Merluccius productus*) using dual beam sonar. M.S. Thesis, University of Washington, 92 pp.

Keiser, R., and J.E. Eherenberg. 1990. An unbiased, stochastic echo-counting model. Rapp. P.-v. Reun. Cons. Int. Explor. Mer 189:65–72.

Kendall, A.W., Jr., and S. Kim. 1989. Buoyancy of walleye pollock (*Theragra chalcogramma*) eggs in relation to water properties and movement in Shelikof Strait, Gulf of Alaska. Can. Spec. Publ. Fish. Aquat. Sci. 108:169–180.

Kim, S., and D.R. Gunderson. 1989. Cohort dynamics of walleye pollock in Shelikof Strait, Gulf of Alaska, during the egg and larval periods. Trans. Am. Fish. Soc. 118:264–273.

Kimura, D.K. 1977. Statistical assessment of the age-length key. J. Fish. Res. Board Can. 34:317–324.

Kimura, D.K. 1989. Variability, tuning, and simulation for the Doubleday-Deriso catch-at-age model. Can. J. Fish. Aquat. Sci. 46: 941–949.

Kimura, D.K., and J.W. Balsiger. 1985. Bootstrap methods for evaluating sablefish pot index surveys. N. Am. J. Fish. Mgt. 5:47–56.

Kjesbu, O.S. 1989. The spawning activity of cod, *Gadus morhua* L. J. Fish. Biol. 34:195–206.

Knudsen, H.P. 1990. The Bergen echo integrator: An introduction. J. Cons. Int. Explor. Mer 47:167–174.

Koeller, P.A. 1981. Distribution and sampling variability in the southern Gulf of St. Lawrence groundfish surveys. Pp. 194–217 in W.G. Doubleday and D. Rivard (eds.), Bottom Trawl Surveys. Can. Spec. Publ. Fish. Aquat. Sci. 58, 273 pp.

Koeller, P.A. 1991. Approaches to improving groundfish survey abundance estimates by controlling the variability of survey gear geometry and performance. J. Northw. Atl. Fish. Sci. 11:51–58.

Koslow, J.A., S. Brault, J. Dugas, and F. Page. 1985. Anatomy of an apparent year-class failure: The early life history of the 1983 Browns Bank haddock *Melanogrammus aeglefinus*. Trans. Am. Fish. Soc. 114:478–489.

Kreiger, K.J. 1993. Distribution and abundance of rockfish determined from a submersible and by bottom trawl. Fish. Bull. 91:(in press).

Kramer, D., and E.H. Ahlstrom. 1968. Distributional atlas of fish larvae in the California Current region: Northern anchovy, *Engraulis mordax* Girard, 1951 through 1965. Cal. Coop. Oceanic Fish. Invest. Atlas 9, 269 pp.

Kuipers, B. 1975. On the efficiency of a two-meter beam trawl for juvenile plaice (*Pleuronectes platessa*). Neth. J. Sea Res. 9:69–85.

Lai, H.L. 1987. Optimum allocation for estimating age composition using age length key. Fish. Bull. 85:179–185.

Lenarz, W.H., and P. B. Adams. 1980. Some statistical considerations of the design of trawl surveys for rockfish (Scorpaenidae). Fish. Bull. 78:659–674.

Lo, N.C.H. 1985. A model for temperature-dependent northern anchovy egg development and an automated procedure for the assignment of age to staged eggs. Pp. 43–50 in R. Lasker (ed.), An Egg Production Method for Estimating Spawning Biomass of Pelagic Fish: Application to the Northern Anchovy, *Engraulis mordax*. NOAA Technical Report NMFS 36.

Lo, N.C.H., J.R. Hunter, H.G. Moser, P.E. Smith, and R.D. Methot. 1992. A daily fecundity reduction method: A new procedure for estimating adult biomass. ICES J. Mar. Res. 49:209–215.

Lo, N.C.H., J.R. Hunter, H.G. Moser, and P.E. Smith. 1993. A daily fecundity reduction method of biomass estimation with application to Dover sole, *Microstomus pacificus*. In J.R. Hunter and N.C.H. Lo (eds.), Ichthyoplankton Methods for Estimating Fish Biomass. Sci. Ser. Los Angeles County Nat. Hist. Museum (in press).

Lockwood, S.J. 1988. The Mackerel: Its Biology, Assessment and the Management of a Fishery. Fishing News Books, Farnham, Surrey, England, 181 pp.

Love, R.H. 1971. Dorsal-aspect target strength of an individual fish. J. Acoust. Soc. Am. 49:816–823.

MacLennan, D.N., and I.G. MacKenzie. 1988. Precision of acoustic fish stock estimates. Can. J. Fish. Aquat. Sci. 45:605–616.

MacLennan, D.N., and J.E. Simmonds. 1992. Fisheries acoustics. Chapman and Hall, London, 325 pp.

Main, J., and G.I. Sangster. 1979. A study of bottom trawling gear on both sand and hard ground. Scottish Fish. Res. Rep. 14. 15 pp.

Main, J., and G.I. Sangster. 1981. A study of the fish capture process in a bottom trawl by direct observations from a towed underwater vehicle. Scottish Fish. Res. Rep. 23. 21 pp.

Matlock, G.C., W.R. Nelson, R.S. Jones, A.W. Green, T.J. Cody, E. Gutherz, and J. Doerzbacher. 1991. Comparison of two techniques for estimating tilefish, yellowedge grouper, and other deepwater fish populations. Fish. Bull. 89:91–99.

Matthews, K.R., and L.J. Richards. 1991. Rockfish (Scorpaenidae) assemblages of trawlable and untrawlable habitats off Vancouver Island, British Columbia. N. Am. J. Fish. Mgt. 11:312–318.

McCall, A.D. 1976. Density dependence of catchability coefficient in the California Pacific sardine, *Sardinops sagax caerulea*, purse seine fishery. Calif. Coop. Ocean. Fish. Invest. Rep., pp. 136–148.

Megrey, B.A. 1989. Review and comparison of age-structured stock assessment models from theoretical and applied points of view. Am. Fish. Soc. Symp. 6:8–48.

Miller, T.J., L.B. Crowder, J.A. Rice, and E.A. Marschall. 1988. Larval size and recruitment mechanisms in fishes: Toward a conceptual framework. Can. J. Fish. Aquat. Sci. 45:1657–1670.

Misund, O.A. 1990. Sonar observations of schooling herring: School dimensions, swimming behavior, and avoidance of vessel and purse seine. Rapp. P.-v. Reun. Cons. Int. Explor. Mer 189:135–146.

Misund, O.A. 1991. Swimming behavior of schools related to fish capture and acoustic abundance estimates. Dr. Philos. Thesis, University of Bergen, 132 pp.

Misund, O.A., and A. Aglen. 1992. Swimming behavior of fish schools in the North Sea during acoustic surveying and pelagic trawl sampling. ICES J. Mar. Res. 49:325–334.

Mitson, R.B. 1983. Fisheries Sonar. Fishing News Books, Farnham, Surrey, England, 287 pp.

Mitson, R.B. 1989. Ship noise related to fisheries research. Proc. Inst. of Acoustics 11(3):61–67.

Morse, W.W. 1989. Catchability, growth, and mortality of larval fishes. Fish. Bull. 87:417–446.

Moser, H.G., and E.H. Ahlstrom. 1985. Staging anchovy eggs. Pp. 37–41 in R. Lasker (ed.), An Egg Production Method for Estimating Spawning Biomass of Pelagic Fish: Application to the Northern Anchovy, *Engraulis mordax*. NOAA Technical Report NMFS 36.

Mundy, P.R. 1979. A quantitative measure of migratory timing illustrated by application to the management of commercial salmon fisheries. Ph.D. Thesis. University of Washington, 87 pp.

Murphy, G.I., and R.I. Clutter. 1972. Sampling anchovy larvae with a plankton purse seine. Fish. Bull. 70:789–798.

Myers, R.A., and W.D. Bowen. 1989. Estimating bias in aerial surveys of harp seal pup production. J. Wildl. Manage. 53: 361–372.

Myhre, R.J. 1969. Gear selection and Pacific halibut. Int. Pac. Hal. Comm. Rep. 51, 35 pp.

Nakashima, B.S. 1990. Escapement from a Diamond IX midwater trawl during acoustic surveys for capelin (*Mallotus villosus*) in the Northwest Atlantic. J. Cons. Int. Explor. Mer 47:76–82.

Nakken, O., and Ø. Ulltang. 1983. A comparison of the reliability of acoustic estimates of fish stock abundances and estimates obtained by other assessment methods in the northeast Atlantic. FAO Fish. Rep. 300:249–260.

Nash, R.D.M., Y. Sun, and C.S. Clay. 1987. High resolution acoustic structure of fish. J. Cons. Int. Explor. Mer 43:23–31.

Nelson, M.O., and E.P. Nunnallee. 1985. Results of acoustic-midwater trawl surveys of spawning pollock in the Shelikof Strait region of the Gulf of Alaska during 1980–81 and 1983–84. NOAA Technical Memo F/NWC-80, pp. 179–206.

Nelson, M.O., and E.P. Nunnallee. 1987. Results of acoustic-trawl surveys for walleye pollock in the Gulf of Alaska in 1986. Pp. 15–38 in R. Major (ed.), Condition of Groundfish Resources in the Gulf of Alaska Region as Assessed in 1986. NOAA Technical Memo. NMFS F/NWC-119, Northwest and Alaska Fisheries Center, Seattle, WA.

Newland, P.L., and C.J. Chapman. 1989. The swimming and orientation behavior of the Norway lobster, *Nephrops norvegicus* (L.), in relation to trawling. Fish. Res. 8:63–80.

Nielsen, L.A. 1983. Variation in the catchability of yellow perch in an otter trawl. Trans. Am. Fish. Soc. 112:53–59.

Noble, R. L. 1970. Evaluation of the Miller high-speed sampler for sampling yellow perch and walleye fry. J. Fish. Res. Bd. Can. 27:1033–1044.

Northcote, T.G., and D.W. Wilkie. 1963. Underwater census of stream fish populations. Trans. Am. Fish. Soc. 92:146–151.

Nunnallee, E.P. 1980. Application of an empirically scaled digital echo integrator for assessment of juvenile sockeye salmon (*Oncorhynchus nerka* Walbaum) populations. Ph.D. thesis, University of Washington, Seattle, WA, 151 pp.

Nunnallee, E. P. 1983. Scaling of an echo integrator using echo counts, and a comparison of acoustic and weir count estimates of a juvenile sockeye salmon population. FAO Fish. Rep. 300:261–268.

Nunnallee, E.P. 1990. An alternative to thresholding during echo-integration data collection. Rapp. P.-v. Reun. Cons. Int. Explor. Mer 189:92–94.

Nunnallee, E.P., and N.J. Williamson. 1989. Results of acoustic–midwater trawl surveys for walleye pollock in Shelikof Strait, Alaska, in 1988. Pp. 225–242 in T.K. Wilderbuer (ed), Condition of Groundfish Resources of the Gulf of Alaska in 1988. NOAA Technical Memo. NMFS F/NWC-165, Northwest and Alaska Fisheries Center, Seattle WA.

Olsen, K. 1990. Fish behavior and acoustic sampling. Rapp. P.-v. Reun. Cons. Int. Explor. Mer 189:147–158.

Olsen, K., J. Angell, F. Petterson, and A. Lovik. 1983. Observed fish reactions to a surveying vessel with special reference to herring, cod, capelin and polar cod. FAO Fish. Rep. 300:131–138

Ona, E. 1989. Trawling noise and fish avoidance related to near-surface trawl sampling. Pp. 169–180 in S. Sundby (ed.), Year Class Variations as Determined from Pre-Recruit Investigations. Inst. Mar. Res., Bergen, 561 pp.

Ona, E. 1990. Physiological factors causing natural variations in acoustic target strength of fish. J. Mar. Biol. Assoc. U.K. 70:107–127.

Ona, E., and O.R. Godø. 1990. Fish reaction to trawling noise: The significance for trawl sampling. Rapp. P.-v. Reun. Cons. Int. Explor. Mer 189:159–166.

Otto, R.S. 1986. Management and assessment of eastern Bering Sea king crab stocks. Can. Spec. Publ. Fish. Aquat. Sci. 92:83–106.

Otto, R.S., J.E. Reeves, and J. Burns. 1977. King and tanner crab research in the eastern Bering Sea, 1977. Int. N. Pac. Fish. Comm. Doc. 2024, 4 pp.

Palsson, W.A. 1984. Egg mortality upon natural and artificial substrata within Washington state spawning grounds of Pacific herring (*Clupea harengus* Pallasi). M.S. Thesis, University of Washington, Seattle, WA, 191 pp.

Parrish, R.H., D.L. Mallicoate, and R.A. Klingbeil. 1986. Age dependent fecundity, number of spawnings per year, sex ratio, and maturation stages in northern anchovy, *Engraulis mordax*. Fish. Bull. 84:503–517.

Pearcy, W.G. 1962. Ecology of an estuarine population of winter flounder *Pseudopleuronectes americanus* Walbaum. III. Distribution, abundance, growth, and production of juveniles; survival of larvae and juveniles. Bull. Bingham Oceanog. Coll. 18:39–64.

Pearcy, W.G., D.L. Stein, M.A. Hixon, E.K. Pikitch, W.H. Barss, and R.M. Starr. 1989. Submersible observations of deep-reef fishes of Heceta Bank, Oregon. Fish. Bull. 87:955–965.

Pedersen, T. 1984. Variation of peak spawning of Arcto-Norwegian cod (*Gadus morhua* L.) during the time period 1929–1982 based on indices estimated from fishery statistics. Pp. 301–316 in E. Dahl, D.S. Danielssen, E. Moksness, and P. Solemdal (eds.), The Propagation of Cod *Gadus morhua* L. Inst. Mar. Res., Flødevigen Biol. Sta.

Pennington, M. 1983. Efficient estimators of abundance for fish and plankton surveys. Biometrics 39:281–286.

Pennington, M. 1986. Some statistical techniques for estimating abundance indices from trawl surveys. Fish. Bull. 84:519–525.

Pennington, M.R., and M.D. Grosslein. 1978. Accuracy of abundance indices based on stratified random trawl surveys. Int. Counc. Explor. Seas. C.M.1978/D:13, 35 pp.

Pennington, M., and J.H. Vølstad. 1991. Optimum size of sampling unit for estimating the density of marine populations. Biometrics 47:717–723.

Peterman, R.M. 1990. Statistical power analysis can improve fisheries research and management. Can. J. Fish. Aquat. Sci. 47:2–15.

Peterson, N.P., and C.J. Cederholm. 1984. A comparison of the removal and mark-recapture methods of population estimation for juvenile coho salmon in a small stream. N. Am. J. Fish. Mgt. 4:99–102.

Picquelle, S. 1985. Sampling requirements for the adult fish survey. Pp. 55–57 in R. Lasker (ed.), An Egg Production Method for Estimating Spawning Biomass of Pelagic Fish: Application to the Northern Anchovy, *Engraulis mordax*. NOAA Technical Report NMFS 36.

Picquelle, S.J., and B.A. Megrey. 1993. A preliminary spawning biomass estimate of walleye pollock, *Theragra chalcogramma*, in the Shelikof Strait, Alaska, based on the annual egg production method. In J.R. Hunter and N.C.H. Lo (eds.), Ichthyoplankton Methods for Estimating Fish Biomass. Sci. Ser. Los Angeles County Nat. Hist. Museum (in press).

Picquelle, S., and G. Stauffer. 1985. Parameter estimation for an egg production method of anchovy biomass assessment. Pp. 7–15 in R. Lasker (ed.), An Egg Production Method for Estimating Spawning Biomass of Pelagic Fish: Application to the Northern Anchovy, *Engraulis mordax*. NOAA Technical Report NMFS 36.

Pitt, T.K., R. Wells, and W.D. McKone. 1981. A critique of research vessel otter trawl surveys by the St. John's research and resource services. Pp. 42–61 in W.G. Doubleday and D. Rivard (eds.), Bottom Trawl Surveys. Can. Spec. Publ. Fish. Aquat. Sci. 58, 273 pp.

Pope, J.G. 1988. Collecting fisheries assessment data. Pp. 63–82 in J.A. Gulland (ed.), Fish Population Dynamics, 2nd ed. John Wiley, New York, 422 pp.

Pope, J.G., and J.G. Shepherd. 1985. A comparison of the performance of various methods for tuning VPAs using effort data. J. Cons. Int. Explor. Mer 42:129–151.

Popper, A.N., and S. Coombs. 1980. Auditory mechanisms in teleost fishes. Am. Sci. 68:429–440.

Press, W.H., B.P. Flannery, S.A. Teukolsky, and W.T. Vetterling. 1989. Numerical recipes; the art of scientific computing (FORTRAN version). Cambridge Univ. Press, 818 pp.

Priede, I.G., and J.J. Watson. 1993. An evaluation of the daily egg production method for estimating biomass of Atlantic mackerel (*Scomber scombrus*). In J.R. Hunter and N.C.H. Lo (eds.), Ichthyoplankton Methods for Estimating Fish Biomass. Sci. Ser. Los Angeles County Nat. Hist. Museum (in press).

Pringle, J.D. 1984. Efficiency estimates for various quadrat sizes used in benthic sampling. Can. J. Fish. Aquat. Sci. 41:1485–1489.

Quinn, T.J., II. 1985. Line transect estimators for schooling populations. Fish. Res. 3:183–199.

Quinn, T.J., II, and V.F. Gallucci. 1980. Parametric models for line-transect estimators of abundance. Ecology 61:293–302.

Quinn, T.J., II, R.B. Deriso, and S.H. Hoag. 1985. Methods of population assessment of Pacific halibut. IPHC Scientific Report 72, 52 pp.

Ralston, S., R.M Gooding, and G.L. Ludwig. 1986. An ecological survey and comparison of bottom fish resource assessments (submersible versus handline fishing) at Johnston atoll. Fish. Bull. 84:141–155.

Reilly, S.B., D.W. Rice, and A.A. Wolman. 1983. Population assessment of the gray whale, *Eschrichtius robustus*, from California shore censuses, 1967–80. Fish. Bull. 81:267–281.

Ricker, W.E. 1975. Computation and interpretation of biological statistics of fish populations. Fish. Res. Board Can. Bull. 191, 382 pp.

Riley, J.D., and J. Corlett. 1965. The numbers of 0-group plaice in Port Erin Bay, 1964–66. Rep. Mar. Biol. Sta. Port Erin. 78:51–56.

Rivard, D. 1989. Overview of the systematic, structural, and sampling errors in cohort analysis. Am. Fish. Soc. Symp. 6:49–65.

Robson, D.S. 1966. Estimation of the relative fishing power of individual ships. ICNAF Res. Bull. 3:5–15.

Rose, C.A., and G.E. Walters. 1990. Trawl width variation during bottom trawl surveys: Causes and consequences. Int. North Pac. Fish. Comm. Bull 50:57–68.

Rose, G.A., and W.C. Leggett. 1988. Hydroacoustic signal classification of fish schools by species. Can. J. Fish. Aquat. Sci. 45:597–604.

Sand, O., and H. E. Karlsen. 1986. Detection of infrasound by the Atlantic cod. J. Exp. Biol. 125:197–204.

Scheaffer, R.L., W. Mendenhall, and L. Ott. 1986. Elementary survey sampling, 3rd ed. Duxbury Press, Boston, 324 pp.

Schill, D.J., and J.S. Griffith. 1984. Use of underwater observations to estimate cutthroat trout abundance in the Yellowstone River. N. Am. J. Fish. Mgt. 4:479–487.

Schwarz, A.L., and G.L. Greer. 1984. Responses of Pacific herring, *Clupea harengus pallasi*, to some underwater sounds. Can. J. Fish. Aquat. Sci. 41:1183–1192.

Schweigert, J.F. 1993. A review and evaluation of methodology for estimating Pacific herring egg deposition. In J.R. Hunter and N.C.H. Lo (eds.), Ichthyoplankton Methods for Estimating Fish Biomass. Sci. Ser. Los Angeles County Nat. Hist. Museum (in press).

Schweigert, J.F., C.W. Haegele, and M. Stocker. 1985. Optimizing sampling design for herring spawn surveys in the Strait of Georgia, B.C. Can. J. Fish. Aquat. Sci. 42:1806–1814.

Schweigert, J.F., C.W. Haegele, and M. Stocker. 1990. Evaluation of sampling strategies for SCUBA surveys to assess spawn deposition by Pacific herring. N. Am. J. Fish. Mgt. 10:185–195.

Seber, G.A.F. 1973. The Estimation of Animal Abundance and Related Parameters. Griffin, London, 506 pp.

Shelton, P.A., M.J. Armstrong, and B.A. Roel. 1993. An overview of the application of the egg production method in the assessment of anchovy in the Southeast Atlantic. In J.R. Hunter and N.C.H. Lo (eds.), Ichthyoplankton Methods for Estimating Fish Biomass. Sci. Ser. Los Angeles County Nat. Hist. Museum (in press).

Shotton, R., and G.P. Bazigos. 1984. Techniques and considerations in the design of acoustic surveys. Rapp. P.-v. Reun. Cons. Int. Explor. Mer 184:34–57.

Simmonds, E.J., and F. Armstrong. 1990. A wideband echo sounder: Measurements on cod, saithe, herring, and mackerel from 27 to 54 kHz. Rapp. P.-v. Reun. Cons. Int. Explor. Mer 189:381–387.

Simmonds, E.J., N.J. Williamson, F. Gerlotto, and A. Aglen. 1992. Acoustic survey design and analysis procedure: A comprehensive review of current practice. ICES Cooperative Research Report 187, 131 pp.

Sinclair, M. 1988. Marine populations; an essay on population regulation and speciation. Wash. Sea Grant, 252 pp.

Sissenwine, M.P. 1978. Using the USA research vessel spring bottom trawl survey as an index of Atlantic mackerel abundance. Int Comm. NW Atl. Fish. Sel. Paper 3:49–55.

Smith, P.E., and S.L. Richardson. 1977. Standard techniques for pelagic fish egg and larva surveys. FAO Fish. Technical Paper 175, 100 pp.

Smith, P.E., W. Flerx, and R.P. Hewitt. 1985. The CalCOFI vertical egg tow (CalVET) net. Pp. 27–32 in R. Lasker (ed.), An Egg Production Method for Estimating Spawning Biomass of Pelagic Fish: Application to the Northern Anchovy, *Engraulis mordax*. NOAA Technical Report NMFS 36.

Smith, P.E., H. Santander, and J. Alheit. 1989. Comparison of the mortality rates of Pacific sardine, *Sardinops sagax*, and Peruvian anchovy, *Engraulis ringens*, eggs off Peru. Fish. Bull. 87:497–508.

Smith, S.J. 1990. Use of statistical models for the estimation of abundance from groundfish trawl survey data. Can. J. Fish. Aquat. Sci. 47:894–903.

Snyder, D.E. 1983. Fish eggs and larvae. Pp. 165–198 in L.A. Nielsen and D.L. Johnson (eds.), Fisheries Techniques. Am. Fish. Soc., Bethesda, MD.

Solemdal, P. 1982. Report of the working group on larval fish ecology. Coun. meeting, Int. Coun. Explor. Sea, 1982 (L:3):71–81 (Mimeo).

Solemdal, P., and B. Ellertsen. 1984. Sampling fish larvae with large pumps; quantitative and qualitative comparisons with traditional gear. Pp. 335–364 in E. Dahl, D.S. Danielssen, E. Moksness, and P. Solemdal (eds.), The Propagation of Cod *Gadus morhua*. L. Inst. Mar. Res., Flødevigen Biol. Sta.

Squire, J.L. 1978. Northern anchovy school shapes as related to problems in school size estimation. Fish. Bull. 76:443–448.

Stauffer, G., and S. Picquelle. 1985. Procedures for sorting, staging, and aging eggs. Pp. 33–35 in R. Lasker (ed.), An Egg Production Method for Estimating Spawning Biomass of Pelagic Fish: Application to the Northern Anchovy, *Engraulis Mordax*. NOAA Technical Report NMFS 36.

Sullivan, P.J. 1991. Stock abundance estimation using depth-dependent trends and spatially correlated variation. Can. J. Fish. Aquat. Sci. 48:1691–1703.

Sundby, S.S., and P. Solemdal. 1984. The egg production of Arcto-Norwegian cod (*Gadus morhua* L.) in the Lofoten area estimated by egg surveys. Pp. 113–135 in O.R. Godø and S. Tilseth (eds.), The Proceedings of the Soviet-Norwegian Symposium on Reproduction and Recruitment of Arctic Cod. Inst. Mar. Res, Bergen, Norway.

Taylor, L.R. 1961. Aggregation, variance and the mean. Nature (London) 189:732–735.

Thorne, R.E. 1983. Application of hydroacoustic assessment techniques to three lakes with contrasting fish distributions. FAO Fish. Rep. 300:269–277.

Thorne, R.E. 1988. An empirical evaluation of the duration-in-beam technique for hydroacoustic estimation. Can. J. Fish. Aquat. Sci. 45:1244–1248.

Thorne, R.E., and G.L. Thomas. 1984. Hydroacoustic measurements of the density and distribution of juvenile sockeye in Tustumena Lake, Alaska, during 1983. Unpubl. Rept. to U.S. Fish and Wildlife Service, 28 pp.

Thorne, R.E., and G.L. Thomas. 1990. Acoustic observations of gas bubble release by Pacific herring (*Clupea harengus pallasi*). Can. J. Aquat. Sci. 47:1920–1928.

Tranter, D.J., and P.E. Smith. 1968. Filtration performance. Pp. 27–56 in D.J. Tranter and J.H. Fraser (eds.), Zooplankton Sampling. UNESCO Monographs on Oceanographic Methodology 2. Paris.

Traynor, J.J., and J.E. Ehrenberg. 1990. Fish and standard-sphere target-strength measurements obtained with a dual-beam and split-beam echo-sounding system. Rapp. P.-v. Reun. Cons. Int. Explor. Mer 189:325–335.

Traynor, J.J., and N.J. Williamson. 1983. Target strength measurements of walleye pollock (*Theragra chalcogramma*) and a simulation study of the dual-beam method. FAO Fish. Rep. 300:112–124.

Traynor, J.J., N. J. Williamson, and W.A. Karp. 1990. A consideration of the accuracy and precision of fish abundance estimates derived from echo-integration surveys. Rapp. P.-v. Reun. Cons. Int. Explor. Mer 189:101–111.

Tsuruta, Y., and K. Hirose. 1989. Internal regulation of reproduction in Japanese anchovy (*Engraulis japonica*) as related to population fluctuation. Can. Spec. Publ. Fish. Aquat. Sci. 108:111–119.

Turnock, B.J., and T.J. Quinn, II. 1991. The effect of responsive movement on abundance estimation using line transect sampling. Biometrics 47:701–715.

Tyler, A.V., and R.S. Dunn. 1976. Ration, growth, and measures of somatic and organ condition in relation to meal frequency in winter flounder, *Pseudopleuronectes americanus*, with hypotheses regarding population homeostasis. J. Fish. Res. Board Can. 33:63–75.

Ulltang, Ø. 1977. Sources of errors in and limitations of virtual population analysis (cohort analysis). J. Cons. Int. Explor. Mer 37:249–260.

Ulltang, Ø. 1980. Factors affecting the reaction of pelagic fish stocks to exploitation and requiring a new approach to assessment and management. Rapp. R.-v. Reun. Cons. Int. Explor. Mer 177:489–504.

Urick, R.J. 1975. Principles of Underwater Sound for Engineers, 2nd Ed. McGraw-Hill, New York, 384 pp.

Uzmann, J.R., R.A. Cooper, R.B. Theroux, and R.L. Wigley. 1977. Synoptic comparison of three sampling techniques for estimating abundance and distribution of selected megafauna: Submersible vs camera sled vs otter trawl. Mar. Fish. Rev. 39:11–19.

Valdes, E.S., P.A. Shelton, M.J. Armstrong, and J.G. Field. 1987. Cannibalism in South African anchovy: Egg mortality and egg consumption rates. S. Afr. J. Mar. Sci. 5:613–622.

van der Veer, H.W. 1986. Immigration, settlement, and density-dependent mortality of a larval and early postlarval 0-group plaice population in the western Wadden Sea. Mar. Ecol. Prog. Ser. 29:223–236.

Venrick, E.L. 1978. Systematic sampling in a planktonic ecosystem. Fish. Bull. 76:617–627.

Wakefield, W.W. 1990. Patterns in the distribution of demersal fishes on the upper continental slope off central California with studies on the role of ontogenetic vertical migration in particle flux. Ph.D. Thesis, University of California, San Diego, CA, 281 pp.

Wakefield, W.W., and A. Genin. 1987. The use of a Canadian (perspective) grid in deep sea photography. Deep Sea Res. 34:469–478.

Wakefield, W.W., and K.L. Smith, Jr. 1990. Ontogenetic vertical migration in *Sebastolobus altivelis* as a mechanism for transport of particulate organic matter at continental slope depths. Limnol. Oceanogr. 35:1314–1328.

Walsh, S.J. 1988. Diel variability in trawl catches of juvenile and adult yellowtail flounder on the Grand Banks and the effect on resource assessment. N. Am. J. Fish. Mgt. 8:373–381.

Walsh, S.J. 1991. Diel variation in availability and vulnerability of fish to a survey trawl. J. Appl. Ichthyol. 7:147–159.

Wardle, C.S. 1977. Effects of size on the swimming speeds of fish. Pp. 299–313 in T.J. Pedley (ed.), Scale Effects in Animal Locomotion. Academic Press, New York.

Wardle, C.S. 1983. Fish reaction to towed fishing gears. Pp. 167–195 in A.G. MacDonald and I.G. Priede (eds.), Experimental Biology at Sea. Academic Press, London.

Wardle, C.S. 1986. Fish behavior and fishing gear. Pp. 463–495 in T.J. Pitcher (ed.), The Behavior of Teleost Fishes. Johns Hopkins University Press, Baltimore, MD.

Wathne, F. 1977. Performance of trawls used in resource assessment. Mar. Fish. Rev. 39:16–23.

West, C.W. 1982. Development and application of bottom trawl instrumentation systems in fisheries resource assessment. Pp. 766–772 in Oceans '82 Conference Record, Marine Technological Soc.

West, C.W. 1985. An examination of variability in the volume swept by bottom trawls used for marine fisheries resource assessment. M.S. Thesis, University of Washington, Seattle, WA, 66 pp.

Westrheim, S.J. 1967. Sampling research trawl catches at sea. J. Fish. Res. Board Can. 24:1187–1202.

Westrheim, S.J. 1976. A further note on sampling research trawl catches at sea. J. Fish. Res. Board Can. 33:1196.

Wilderbuer, T.K. 1988. Fishing power standardization in multi-vessel, multi-gear, resource assessment surveys. M.S. Thesis, University of Washington, Seattle, WA, 171 pp.

Wilkins, M.E. 1986. Development and evaluation of methodologies for assessing and monitoring the abundance of widow rockfish, *Sebastes entomelas*. Fish. Bull. 84:287–310.

Williamson, N.J. 1982. Cluster sampling estimation of the variance of abundance estimates derived from quantitative echo sounder surveys. Can. J. Fish. Aquat. Sci. 39:229–231.

Witthames, P.R., and M. Greer Walker. 1987. An automated method for counting and sizing fish eggs. J. Fish Biol. 30:225–235.

Zhang, C.I., and P.J. Sullivan. 1988. Biomass-based cohort analysis that incorporates growth. Trans. Am. Fish. Soc. 117:180–189.

Zjilstra, J.J., R. Dapper, and J.J. Witte. 1982. Settlement, growth and mortality of post-larval plaice (*Pleuronectes platessa*) in the western Wadden Sea. Neth. J. Sea Res. 15:250–272.

AUTHOR INDEX

SUBJECT INDEX